Life in the Sea

The Coast

Pam Walker and
Elaine Wood

Facts On File, Inc.

The Coast

Copyright © 2005 by Pam Walker and Elaine Wood

All rights reserved. No part of this book may be reproduced or utilized in any form or by any means, electronic or mechanical, including photocopying, recording, or by any information storage or retrieval systems, without permission in writing from the publisher. For information contact:

Facts On File, Inc.
132 West 31st Street
New York NY 10001

Library of Congress Cataloging-in-Publication Data
Walker, Pam, 1958–
 The coast/ Pam Walker and Elaine Wood.
  p. cm. — (Life in the sea)
 Includes bibliographical references and index.
 ISBN 0-8160-5701-X (hardcover)
 1. Coastal ecology—Juvenile literature. 2. Coasts—Juvenile literature.
  I. Wood, Elaine, 1950– II. Title.
  QH541.5.C65W35 2005
  578.769'9—dc22   2004024223

Facts On File books are available at special discounts when purchased in bulk quantities for businesses, associations, institutions, or sales promotions. Please call our Special Sales Department in New York at
(212) 967-8800 or (800) 322-8755.

You can find Facts On File on the World Wide Web at
http://www.factsonfile.com

Text and cover design by Dorothy M. Preston
Illustrations by Dale Williams, Sholto Ainslie, and Dale Dyer

Printed in the United States of America

VB FOF 10 9 8 7 6 5 4 3 2 1

This book is printed on acid-free paper.

Contents

Preface vii
Acknowledgments viii
Introduction ix

1. Physical Aspects: Coastal Waters, Waves, and Substrates 1
Greenhouse Gases 2
Types of Coasts 5
Features of Coasts 8
Science of Coastal Waters 10
Chemical and Physical Characteristics of Water 12
Substrates Along the Coast 14
Marine Processes: Tides, Waves, Wind 15
Tides 16
Coastal Habitats 19
Conclusion 20

2. Microbes and Plants: Bacteria, Protists, Plants, and Fungi Along the Coast 22
Food Chains and Photosynthesis 23
Monerans 24
Ancient Cyanobacteria 25
Heterotrophic Bacteria 26
Kingdoms of Living Things 26
Protists and Fungi 28
Advantages of Sexual Reproduction 30

Plants of the Coast 32
      Light and Algal Coloration 33
      Green Algae 33
      Differences in Terrestrial and Aquatic Plants 34
      Brown Algae 36
      Plant Defenses 38
      Red Algae 38
      Sea Grasses 39
   Conclusion 41

3. Sponges, Cnidarians, and Worms: Simple Invertebrates in Coastal Waters 42
   Sponges 42
      Body Symmetry 46
   Cnidarians 48
      Spawning and Brooding 52
   Worms 53
      Worm Comparisons 56
   Conclusion 59

4. Mollusks, Arthropods, and Echinoderms: Complex Coastal Animals 62
   Mollusks 63
   Arthropods 69
      Advantages and Disadvantages of an Exoskeleton 70
      Crustaceans 71
      Sea Spiders and Horseshoe Crabs 76
   Echinoderms 77
   Conclusion 81

5. Coastal Fish: Life in Shallow Seawater 83
   Sculpins 83
      Bony Fish Anatomy 86

Gobies 88
   Colorization 88
Blennies 90
   The Skin and Senses of Fish 92
Gunnels 92
   Water Balance 94
Clingfish 95
   Territoriality 96
Conclusion 96

6. Reptiles, Birds, and Mammals: Vertebrates at the Edge of the Ocean 98

   Body Temperature 98
Marine Reptiles 99
   Marine Reptile Anatomy 100
Seabirds 103
   Marine Bird Anatomy 104
Mammals 108
   Marine Mammal Anatomy 109
Conclusion 113

7. Change Is Constant 114

Forces That Influence the Coast .115
The Impact of Humans 116

Glossary 119
Further Reading and Web Sites .125
Index 129

Preface

Life first appeared on Earth in the oceans, about 3.5 billion years ago. Today these immense bodies of water still hold the greatest diversity of living things on the planet. The sheer size and wealth of the oceans are startling. They cover two-thirds of the Earth's surface and make up the largest habitat in this solar system. This immense underwater world is a fascinating realm that captures the imaginations of people everywhere.

Even though the sea is a powerful and immense system, people love it. Nationwide, more than half of the population lives near one of the coasts, and the popularity of the seashore as a home or place of recreation continues to grow. Increasing interest in the sea environment and the singular organisms it conceals is swelling the ranks of marine aquarium hobbyists, scuba divers, and deep-sea fishermen. In schools and universities across the United States, marine science is working its way into the science curriculum as one of the foundation sciences.

The purpose of this book is to foster the natural fascination that people feel for the ocean and its living things. As a part of the set entitled Life in the Sea, this book aims to give readers a glimpse of some of the wonders of life that are hidden beneath the waves and to raise awareness of the relationships that people around the world have with the ocean.

This book also presents an opportunity to consider the ways that humans affect the oceans. At no time in the past have world citizens been so poised to impact the future of the planet. Once considered an endless and resilient resource, the ocean is now being recognized as a fragile system in danger of overuse and neglect. As knowledge and understanding about the ocean's importance grow, citizens all over the world can participate in positively changing the ways that life on land interacts with life in the sea.

Acknowledgments

This opportunity to study and research ocean life has reminded both of us of our past love affairs with the sea. Like many families, ours took annual summer jaunts to the beach, where we got our earliest gulps of salt water and fingered our first sand dollars. As sea-loving children, both of us grew into young women who aspired to be marine biologists, dreaming of exciting careers spent nursing wounded seals, surveying the dark abyss, or discovering previously unknown species. After years of teaching school, these dreams gave way to the reality that we would not spend our careers working with sea creatures, as we had hoped. But time and distance never diminished our love and respect for the oceans and their residents.

We are thrilled to have the chance to use our own experiences and appreciation of the sea as platforms from which to develop these books on ocean life. Our thanks go to Frank K. Darmstadt, executive editor at Facts On File, for this enjoyable opportunity. He has guided us through the process with patience, which we greatly appreciate. Frank's skills are responsible for the book's tone and focus. Our appreciation also goes to Katy Barnhart for her copyediting expertise.

Special notes of appreciation go to several individuals whose expertise made this book possible. Audrey McGhee proofread and corrected pages at all times of the day or night. Diane Kit Moser, Ray Spangenburg, and Bobbi McCutcheon, successful and seasoned authors, mentored us on techniques for finding appropriate photographs. We appreciate the help of these generous and talented people.

Introduction

No other part of the ocean is easier to get to or more often visited than its coast. The intertidal zone, that space of coast between where the highest high tide rises and the lowest low tide reaches, has been explored by millions of bare feet, probing fingers, and curious eyes. In one sense, people are more acquainted with the intertidal zone than they are with any other part of the ocean. Acquaintance is the first step toward appreciating and understanding this complex ecosystem. The next step is gaining information.

The Coast is one volume in a set of books by Facts On File entitled Life in the Sea, a group of texts that examine the biology of the major regions of the ocean. The focus of *The Coast* is on the adaptations and relationships of organisms that live in the zone between the low-tide and high-tide marks.

Chapter 1 takes a close look at the aspects of geology, physical science, and biology that shape life along the shoreline. Beginning with a review of the history of shorelines across geologic time, this chapter pays particular attention to the geologic forces that are responsible for the current structure of shores. Chapter 1 also examines the way in which the presence, or absence, of water creates distinct zones in which organisms can be found. The upper littoral zone is the one that is most distant from the water. In it, organisms are more terrestrial than marine because their only source of water is the surf's splash. Moving seaward, the next zone is the mid-littoral, an area that is covered with water once or twice daily. Inhabitants are marine organisms whose bodies are able to conserve moisture when the tide is out. The lower littoral zone is underwater most of the time, suffering exposure to air only during extremely low tides. This section can support a greater diversity of organisms than the other two areas.

Chapter 2 introduces some of the microorganisms, fungi, and plants that live in the intertidal zones. Cyanobacteria (very primitive green cells) and diatoms (more advanced single-celled photosynthesizers) are the primary producers in this system. Both types of organisms are capable of using the Sun's energy to make glucose and other organic molecules that are essential to life. Other producers include the shoreline macroalgae, such as *Ulva* and *Porphyra,* that serve as food for many organisms. Green organisms support a variety of food chains that are essential for animal life in the intertidal ecosystem. Fungi and heterotrophic bacteria decompose dead plant and animal matter that accumulates there and by doing so provide food for an entirely different group of animals than those supported by plants.

Chapters 3 and 4 investigate representatives of the invertebrate groups in the intertidal zone, the small animals. These organisms include sponges, cnidarians, worms, mollusks, crustaceans, and echinoderms. In the coastal food chain, these invertebrates feed on plants and animals and serve as food for larger organisms. Most are protected by structures such as shells or spines or by toxic chemicals.

Sponges are such simple animals that for centuries they were mistakenly classified as plants. Looking like anything from crusts or vases to fingers or antlers, sponges also vary in color from dull to vibrant. Many of them serve as homes to small animals. Alongside intertidal sponges are the cnidarians (also called coelenterates), which include anemones, hydrozoans, and jellyfish. Although anemones and hydrozoans are common in tidal pools, the only jellyfish native to intertidal water is the stalked jellyfish. All cnidarians share the same saclike body plan, with rings of armed tentacles that capture food and repel predators.

Less obvious, but just as numerous, are flatworms and segmented worms. Some species are free living, but many inhabit tubes just under the surface of the sand or mud. Crawling slowly among theses relatively simple animals are the larger arthropods—mollusks and echinoderms.

Coastal vertebrates, animals that have backbones, are easier to spot than invertebrates. Chapters 5 and 6 explore the fish, reptiles, birds, and mammals that make their homes in or near the intertidal waters. Coastal fish are small, with relatively large heads and long, tapered bodies. Some can jump from one tide pool to another, a technique that expands their feeding range and increases their chances of finding mates. Others are outfitted with suckerlike structures that enable them to cling to rocks when energetic seawater threatens to wash them away. A few are even short-term air breathers, an unusual adaptation in fish. By crawling outside their tide pools when oxygen levels are extremely low, they can survive until fresh, oxygenated water returns in the next high tide.

The only seaside reptile is the marine iguana, a large lizard found on the Galápagos Islands. Shorebirds, however, are numerous and include gulls, plovers, oystercatchers, and sandpipers. These familiar birds wade or soar over the shallow intertidal waters, looking for small invertebrates. Most have bills that can probe into the sand or between the rocks to pluck tasty morsels from their hiding spots. The feet of many shorebirds are lobed to keep them from sinking into wet soil.

Seashore mammals belong to the fin-flippered group, which includes sea lions and walruses, large animals whose back legs are fused and front legs are modified to form fins. Though slow and clumsy on land, in the water they are quick, graceful acrobats that can go on extended dives. Seals stay warm because they have a thick layer of insulating blubber under their skins. These large animals prey on fish and a variety of marine invertebrates.

Faced with challenges not found in any other part of the ocean, organisms on the coast are extremely well adapted for their environments. Each species plays a vital role in the cycle of life and death that keeps the seaside ecosystem running smoothly. Learning more about the intertidal ecosystems and the organisms in them helps each of us preserve these dynamic windows into the ocean, as discussed in chapter 7.

1

Physical Aspects
Coastal Waters, Waves, and Substrates

The shoreline, the area where the land meets the sea, is the most familiar part of the ocean for nearly everyone. It is also the place where many people have their first sea experiences. The shore attracts visitors for a variety of reasons. For millions, it is a special place of rest or play, the destination of choice on vacations and holidays. Some prefer to make their permanent homes there, living within the sound of the surf and the view of the open water. Others depend on the shore area for their livelihood, using it as a base of operations for work at sea.

Fig. 1.1 The coast of Oregon is a rocky shoreline. (Courtesy of Rear Admiral Harley D. Nygren, NOAA Corps, Ret.)

Greenhouse Gases

Carbon dioxide is one of several so-called greenhouse gases that form an invisible layer around the Earth. As shown in Figure 1.2, greenhouse gases trap the Sun's heat near the Earth's surface, very much like the windows in a greenhouse hold in heat from the Sun. The greenhouse gases are one of the reasons that temperatures on Earth's surface are warm enough to support life. If they did not exist in the atmosphere, most of the Sun's radiant energy would bounce off the Earth's surface and return to space.

The layer of greenhouse gases is changing, however, and this change has many scientists worried. By burning fossil fuels in homes, cars, and industries, people all over the world are constantly adding carbon dioxide to the air, widening the belt of greenhouse gases. Many environmentalists fear that the rising levels of carbon dioxide in the air are warming the Earth's surface abnormally, a phenomenon known as global warming.

Research indicates that some warming has already taken place in the air and in the ocean. The effects of this warming include less snow cover each winter, a retreat of mountain glaciers, and changes in global weather patterns. Experts fear that continued warming could damage the balance of life on Earth. Some predict far-reaching results, including changes in climates, melting of glacial ice, and damage to the coral reefs.

Fig. 1.2 Carbon dioxide is one of the greenhouse gases in the atmosphere that traps heat close to the surface of the Earth.

Physical Aspects 3

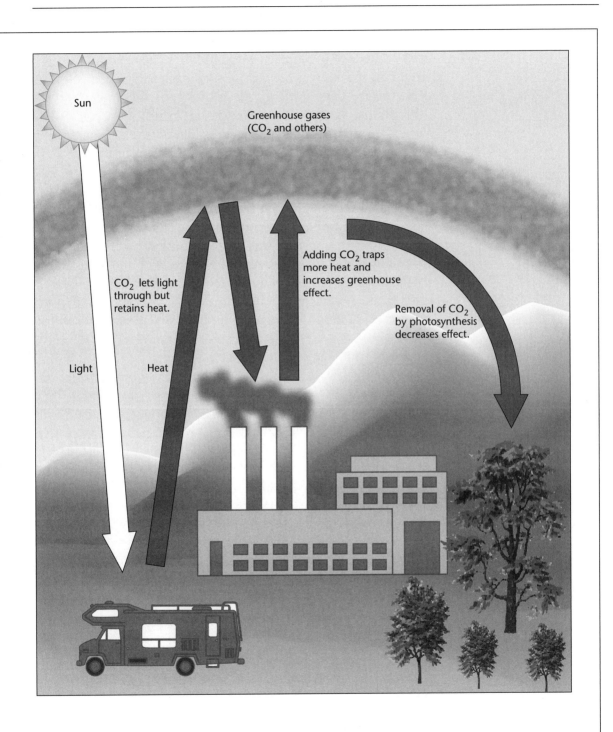

The world's shorelines are far reaching, measuring a total of more than 273,000 miles (440,000 km). They extend through all weather zones of the world, including the extremes of the freezing ice caps and the steaming tropics. Some areas are rocky, such as the Oregon coast shown in Figure 1.1, while others are sandy. A wide variety of habitats, each containing an incredible diversity of organisms, is incorporated in these borders between land and sea.

The shore is part of a larger zone referred to as the coast, the entire area that is affected by the sea. The coast includes the familiar sand and surf, as well as mud flats, tide pools, and marshes. The coast begins at the point where waves start breaking as they roll in toward the shore, and it extends to the farthest reaches of waves and tides on land. In some localities, the distance between the first breaking waves and the highest tides is just a matter of meters; in others, it encompasses miles.

No other part of the Earth is more dynamic than the coast. Some of the changes it experiences are both quick and extreme, like those caused by a storm. Others, such as its gentle reshaping by wind and waves, occur so slowly that they can only be observed over decades. Yet, when viewed over eons, these slow changes are as significant as those carried out by the most severe storms.

Throughout the Earth's history, the locations and characteristics of coasts correspond to the amount of water in the oceans. Although unnoticeable in a single human lifetime, the volume of the world's oceans has varied dramatically over the life of the Earth. Two of the primary factors that determine ocean volume are the amount of ice in the ocean and the size of the ocean basin.

Ice plays a role in ocean volume because freezing effectively removes water from the ocean, causing water levels to drop and the coastlines to widen. During each major ice age, the ocean contained less water than it does at the present time. When so much water "disappeared" into ice, continents were wider and shorelines extended far past their present-day borders.

The container that holds the oceans' water is called the ocean basin. As the rest of the Earth's crust, the basin is made of movable plates that can shift positions, much like a bowl whose size is expandable. When the size of the basin changes, so does the level of water in that basin.

The plates that make up the Earth's crust are always moving, so the outer shape is constantly being modified. In the oceans, there are places where hot lava bubbles up from the mantle to the Earth's surface. When lava emerges from the interior of the Earth, it pushes the crustal plates apart, increasing the size of the ocean basin. As the ocean container gets bigger, the level of water in the sea drops.

Over most of the last 3,500 years, ocean levels have remained stable. However, in recent years the ocean's elevation has increased slightly. Many scientists feel that this latest change is not related to changes in the crust but to increased global warming, a gradual rise in the temperatures on the surface caused by an enlarging layer of greenhouse gases in the atmosphere.

Types of Coasts

Coastal areas exist in a wide variety of shapes and forms. For this reason, scientists have found it useful to divide them into subgroups or types. One way of sorting coastlines is by when and how they were formed.

All landforms, whether they are coasts or mountains, are formed and changed by geological processes. Coastlines can be divided into two large groups, based on whether their traits were primarily defined by land processes or by sea processes. Those sculpted by land processes are called primary coasts, and the ones that have been shaped by the ocean are called secondary coasts.

The types of land processes that shape the appearance of a primary coast include precipitation (rain, snow, sleet, and hail), erosion, and deposition of sediments by wind and water. On a geologic time scale, primary coasts are fairly young and have been in very much the same condition since after the last ice age, 6,500 years ago. In this short stretch of geologic time, the ocean has not had time to alter them.

The soil on a primary coast was once a part of the land. In some cases, the soil was deposited on the coast by wind or running water. In other cases, slow-moving glaciers pushed soil to the coast. A few coasts are made of soil derived from volcanic activity. Others are made of large chunks of land that were displaced by earthquakes or by movements along fault lines.

In the last ice age, when much of the water of the world's oceans was frozen as ice and sea levels were lower than they are now, the coastlines extended past their present positions. Then, as now, many rivers flowed to the sea, cutting deep, V-shaped valleys as they went. Thousands of years later, when much of the ice caps melted and sea levels rose, water filled in, or "drowned," these river troughs. The type of coast that is dominated by an old river valley is called a drowned river, or ria coast (after the Spanish term *ría*, which means "estuary"). The Chesapeake Bay is a good example of this type of coastline.

Land processes that push or deposit soil out into the ocean form another type of primary coastline, the built-out coast. Deltas are built-out coasts created from deposits of sediment. The Earth's rivers move an incredible amount of soil, delivering as much as 530 tons to sea every second. At the mouth of a river, water slows down and much of the sediment suspended by the water's high energy begins to settle. If the sea at a river's mouth is energetic and deep, sediment is quickly swept away; however, if the river empties into a protected part of the sea where ocean forces cannot disperse it, then soil builds up to form a delta.

At the point where the Mississippi River meets the Gulf of Mexico, one of the largest deltas in the United States is still growing. Other large deltas include those formed in the Mediterranean Sea by the Nile, Rhône, Po, and Ebro Rivers, as well as those created where the Ganges-Brahmaputra River joins the Bay of Bengal, and where rivers empty into the South China Sea.

Glaciers are another land-based force of nature that can create built-out coasts. During the last ice age, glaciers slid across the continents on the way to the sea, gouging out deep valleys and pushing tons of soil and rock in front of them. When the glaciers melted, the soil that they were moving was left behind, forming ridges called moraines. Moraines left at these extended coasts became part of the coastline. Later, when sea levels rose, some of these piles of rubble were completely or partially surrounded with water. Long Island, off the New York and Connecticut coasts, and Cape Cod, Massachusetts, are moraine islands. Martha's Vineyard and Nantucket are more distant reaches of glacial deposits on the northeastern coast of the United States.

Volcanoes are responsible for creating several built-out coasts. In some parts of the ocean, the eruption of undersea volcanoes erected mountains of lava that eventually reached the water's surface and formed islands. As these mountains grew, their coastlines were dominated by lava flows. The Hawaiian Islands were formed from volcanic activity.

Primary coasts are also formed by shifts in the Earth's crust. When plates of the crust change positions, they can create large tears or splits called faults. If a fault forms at the coast, seawater rushes into it and creates a fault bay. The Gulf of California, also called the Sea of Cortés, lies between Baja California and mainland Mexico. At one time, Baja California was part of the North American continent. When the crustal plates slid horizontally past each other, a bit of land, the area now known as Baja California, was ripped away from the continent.

Fault bays are formed by horizontal movement of Earth's plates, but plates can also move up and down. If the seabed moves down and the continental mass remains in place, tall cliffs form along the coast. On the other hand, if the continent shifts upward, places that were once under water are suddenly exposed. This kind of movement pushed up much of the seafloor in Prince William Sound, Alaska, after an earthquake on March 27, 1964.

Secondary coasts are areas that have been changed by marine processes. Like primary coasts, they were originally formed by processes on land, but they have been around longer than primary coasts, long enough for their appearance to be influenced by action of the sea.

Water, waves, and currents are some of the sea forces that mold secondary coasts. Water is a great solvent that dissolves minerals in rock and soil. In addition, ocean water contains particles such as sand, small stones, and gravel that act like sandblasters, eroding structures and changing the coastline.

The rate at which the ocean erodes the coast is determined by factors such as the composition of the rocks making up the coast, the types of soil and rock carried by the water, and the energy level of water. Coasts made of hard rock like granite erode very slowly. The coasts of Maine are granite, and they recede only one or two inches (2–5 cm) every 10 years. However, coasts made up of sand or sandstone are much softer, so they can change dramatically, sometimes disappearing at the rate of several yards per year. The North Sea cliffs in England, which are made of soft stone, were worn back more than 36 feet (11 m) during one severe storm.

Coasts pounded by high-energy water endure powerful waves and frequent storms. The coasts of the eastern United States and Canada, as well as the southern tips of South America and South Africa, are high-energy areas. Low-energy coasts, where few big waves and storms appear, are often in protected locations such as gulfs.

Features of Coasts

Ocean forces create a variety of features on secondary coasts. The constant erosion caused by waves pounding on the shore carves out sea cliffs and caves. Just off the coast, the same wave action sculpts natural arches or flat platforms. In places where the underwater slope of the seafloor is not steep, waves and tides can deposit sediment and build an area of loose particles called a beach. In the United States, about 30 percent of the coastlines have beaches.

Beaches are subject to seasonal changes. The low, gentle waves of summer bring sand to a beach. During the winter, storms create higher, stronger waves that carry away sand. In the Northern Hemisphere, the most severe wave action starts in December but slows significantly by April.

Some of the features of beaches are created by wave action. An accumulation of sediment that is deposited parallel to shore forms a section called the berm. The berm marks the upper limit of sand deposition by waves. The top of the berm, the berm crest, is usually the highest place on the beach. The backshore, an area made up of sand that is deposited in dunes, is on the landside of the berm crest. On the seaside, the area between the berm crest and sea is the foreshore, an active stretch that is constantly washed with waves.

The sand on a beach displays several interesting features, as seen in the upper color insert on page C-1. Ripples are the marks made in sand by waves that rush onto the shore. Rills are small, branched depressions in the sand that drain water back toward the ocean. Diamond-shaped deposits of silt are backwash marks, places where the shells of animals interfere with the normal backwash of water. Regularly spaced, crescent-shaped depressions along the sand are called cusps. No one knows for sure how cusps form, but many believe them to be due to irregularities along the beach that are enlarged by swash, the water that runs off the beach after a wave breaks.

Some shores are bordered by barrier islands, exposed sandbars that run parallel to the shore. Worldwide, about 13 percent of the coasts have barrier islands. These protective, sandy walls form in one of three ways. Some are the result of sediment deposits just offshore, like the islands off the coasts of Alabama and Mississippi. Others were ancient sand dunes that formed on the extended beaches of the last ice age. When glaciers melted and sea levels rose, these dunes were surrounded by water. Most of the islands off the coast of the southeastern United States, including the Outer Banks of North Carolina and Georgia's Tybee Island, formed in this way. Another type, called a sea island, was actually part of the mainland that remained exposed when the sea level rose. Sea islands, like Cumberland and Hilton Head off the coast of Georgia, are not as sandy as dunelike barrier islands.

A few secondary coasts owe their characteristics to marine organisms rather than to the sea's physical processes. Both plants and animals can add their own customized touches to a coast. In the tropics, small anemone-like sea animals called corals build extensive reefs along the coasts. Coral animals surround themselves with a hard skeleton of calcium. When a coral animal dies, its skeleton becomes part of an ever-growing underwater skyscraper made of living organisms atop the skeletons of dead ones. The Florida Keys, a string of islands off the tip of Florida, have coral reef coasts.

One of the primary architects in the plant kingdom is the mangrove tree, a woody plant that can grow in salt water. Mangroves send out extensive, aboveground roots that trap and hold sediment suspended in seawater, retaining enough soil to extend the size of a landform. An assortment of living things finds the mangrove root community to be an ideal home and a rich source of food. Mangrove coasts are common in Florida, northern Australia, and in the Bay of Bengal in the East Indies.

Science of Coastal Waters

The marine environment is, to a great extent, defined by the physical and chemical characteristics of ocean water. These characteristics and the limits to which they extend are very important in determining what kinds of organisms can live in a region. Both chemical and physical factors, which include salinity, levels of dissolved gases, density, and temperature, are more extreme in shallow waters than in other parts of the ocean.

Anyone who has tasted seawater knows that it is salty. Although salinity is fairly constant throughout most of the ocean, it is exceptionally variable near the shore. Salinity refers to the amount of dissolved minerals, or salts, in water. The average salinity of ocean water is 35 parts per thousand, abbreviated as 35‰ (per ml). (The symbol ‰ is similar to percent but refers to parts per *thousand* instead of parts per *hundred*.)

Salts in ocean water come from dissolved solids that originate on the land. The action of weathering slowly dissolves rocks and minerals, which are carried to the ocean by water in streams, creeks, and rivers. A small percentage of ocean minerals also come from the atmosphere and from the Earth's interior.

Most of the minerals dissolved in water form sodium ions and chloride ions. Ions are charged particles created when minerals break down and dissolve in water. Some of the other ions that find their way to the ocean are those of sulfate, magnesium, calcium, and potassium.

The salinity of water can be affected by a lot of factors. Anything that removes water from the ocean causes the salinity to increase. When water evaporates or freezes, it is removed from the ocean, leaving behind the salts in it. In places where water evaporates from slow-moving or stagnant pools of salt water, salinity tends to be higher than in the rest of the ocean. On the other hand, any factor that adds water to the oceans decreases the salinity. Salinity is relatively low in areas where freshwater flows into the ocean, such as near the mouths of rivers.

Along the coast, salinity levels can vary extremely. In areas where a river enters the ocean, or places where there is a lot of precipitation, salinity drops below average. For example, the water at the mouth of the Amazon River, where it runs into the Atlantic Ocean, has a salinity that is 25 percent lower than surrounding water. The same thing happens where rivers empty into bays and harbors.

In hot climates, evaporation rates are high and ocean salinity ranges on the upper end of the scale. In the Red Sea and Persian Gulf, salinity may reach 42 percent. Salinity is also elevated in places where little or no new freshwater enters the system, or where water is trapped without a natural outlet. In the Dead Sea, water flows in from the Jordan River, but it has no path by which to leave the system.

In any body of salt water, the freshest water is the top layer and saltiness increases with depth. At 130 feet (40 m) the salinity of the Dead Sea is 300 parts per million (ppm), about 10 times greater than the ocean's. The only organisms that can survive this extreme environment are a few species of bacteria.

Just as there are gases in the atmosphere surrounding the Earth, there are gases in Earth's waters. Oxygen and carbon dioxide are two gases that play critical roles in life in the sea. Like the creatures that live in an ocean of air, marine organisms require gases to survive. The levels of dissolved gases in coastal waters can vary dramatically. Generally, water in the surf is well mixed with both oxygen and carbon dioxide. In shallow tide pools, where water is less energetic, the levels of oxygen can drop quickly.

The amount of sunlight in a system affects how much photosynthesis can occur there. Since one of the raw materials of photosynthesis is carbon dioxide, and one of the by-products is oxygen, the rate at which this reaction occurs also affects

Chemical and Physical Characteristics of Water

Water is one of the most widespread materials on this planet. Water fills the oceans, sculpts the land, and is a primary component in all living things. For all of its commonness, water is a very unusual molecule whose unique qualities are due to its physical structure.

Water is a compound made up of three atoms: two hydrogen atoms and one oxygen atom. The way these three atoms bond causes one end of the resulting molecule to have a slightly negative charge, and the other end a slightly positive charge. For this reason water is described as a polar molecule.

The positive end of one water molecule is attracted to the negative end of another water molecule. When two oppositely charged ends of water molecules get close enough to each other, a bond forms between them. This kind of bond is a hydrogen bond. Every water molecule can form hydrogen bonds with other water molecules. Even though hydrogen bonds are weaker than the bonds that hold together the atoms within a water molecule, they are strong enough to affect the nature of water and give this unusual liquid some unique characteristics.

Water is the only substance on Earth that exists in all three states of matter: solid, liquid, and gas. Because hydrogen bonds are relatively strong, a lot of energy is needed to separate water molecules from one another. That is why water can absorb more heat than any other material before

levels of dissolved gases. On land, most plants live above the soil in easy access of sunlight. But in the water, light penetration is limited by depth and by cloudiness.

Temperature plays critically important roles in both living and nonliving systems. Temperature, a measure of the amount of heat in a system, determines the rate at which chemical reactions take place. Up to a point, the more heat there is in a system, the faster the molecules in that system move. Therefore, when temperatures are high, molecules are more active and more likely to encounter one another. That is why warm temperatures increase the rates of chemical reactions. In living things, the rate at which chemical reactions occur is referred to as the metabolic rate. As temperature increases, so does metabolic rate, doubling with a change of 18°F (10°C). However, too much heat distorts the structures

its temperature increases and before it changes from one state to another.

Since water molecules stick to one another, liquid water has a lot of surface tension. Surface tension is a measure of how easy or difficult it is to break the surface of a liquid. These hydrogen bonds give water's surface a weak, membrane-like quality that affects the way water forms waves and currents. The surface tension of water also impacts the organisms that live in the water column, water below the surface, as well as those on its surface.

Atmospheric gases, such as oxygen and carbon dioxide, are capable of dissolving in water, but not all gases dissolve with the same ease. Carbon dioxide dissolves more easily than oxygen, and there is always plenty of carbon dioxide in seawater. On the other hand, water holds only $\frac{1}{100}$ the volume of oxygen found in the atmosphere. Low oxygen levels in water can limit the number and types of organisms that live there. The concentration of dissolved gases is affected by temperature. Gases dissolve more easily in cold water than in warm, so cold water is richer in oxygen and carbon dioxide than warm water. Gases are also more likely to dissolve in shallow water than deep. In shallow water, oxygen gas from the atmosphere is mixed with water by winds and waves. In addition, plants, which produce oxygen gas in the process of photosynthesis, are found in shallow water.

of molecules in living things. The absence of heat slows molecular activity so much that molecules cannot interact and reactions do not occur.

The temperatures of ocean environments vary with geographic location, water depth, and the seasons. On average, the ocean's temperature is only a few degrees above freezing. The warmest marine waters are found at the surface and in coastal areas. The coolest are found near the poles, in the open ocean, and near the ocean floor.

Both temperature and salinity affect another important physical characteristic of seawater: density. Density is a property of matter that refers to its mass per unit volume. A cup of seawater has a higher density than a cup of freshwater because of the added mass of the dissolved salts. As the salinity of water increases, so does its density.

Temperature influences density because it affects the volume of water. As temperature increases, water expands and takes up more space. Since the mass of warm water is spread over a larger volume than the mass of a similar amount of cool water, warm water has a lower density.

Working together, salinity and temperature regulate water's density. Density of water is an important factor because it determines where water will be located in the water column, water below the surface. Since denser water sinks below less dense water, both very salty and extremely cold waters move to the lowest levels of water columns. Cold, salty water is the densest kind. Warm seawater that is mixed with some freshwater is not dense, so it rides on top of the water column. In many coastal areas, the influx of freshwater from streams and rivers reduces the density of the ocean water.

Substrates Along the Coast

The chemistry of seawater is only one aspect of a marine environment. Substrates also play important roles because they serve as homes to many organisms. The substrate of a marine environment refers to the material that makes up the bottom or floor of the area, and it can be one of three types: loose and sandy, muddy and compact, or hard and rocky.

Hard, rocky substrates are characteristic of tide pools, little puddles of seawater that are left behind when high tide recedes. The conditions in tide pools are extremely variable, and the organisms that live there are specialized to these conditions. Tide pools are dynamic environments that are alternately wet and dry. Inhabitants are adapted to cling to rocky surfaces when water rushes into tide pools, and to conserve moisture when water levels are low.

The sandy beach provides an entirely different substrate. Just as in tide pools, water is intermittent on beaches. However, the organisms on beaches do not have anything to cling to, so adaptations for holding are not very useful. Most beach organisms are burrowers who dig into the sand to keep from washing out to sea. Because they cannot maintain their positions in shifting sands, there are not many multicelled photosynthetic organisms, like grasses or shrubs, inhabiting sandy areas. However, large populations of photosynthetic single-celled organisms live on and among the sand grains.

Mud flats are subject to the coming and going of tides, but unlike beaches and tide pools, they are protected from waves. Water cannot drain through muddy soil like it does on beaches where soil particles are loosely packed. Mud particles are small and tightly packed, so they hold water in place. When water occupies the spaces between mud particles, there is no room for oxygen. As a result, mud flats are anaerobic environments that support only organisms that can survive in low or no oxygen.

Marine Processes: Tides, Waves, Wind

Every ocean visitor has seen waves, ridges of water that seem to be traveling across the ocean's surface. Despite the illusion, water does not really travel in a wave. The only thing that travels in a wave is energy; the water just moves up and down. The original energy that starts waves comes from several sources, but the most common one is wind. Others include

Fig. 1.3 The highest part of the wave is called the crest, and the lowest part, the trough. Individual water particles in a wave move in a circular path called an orbit.

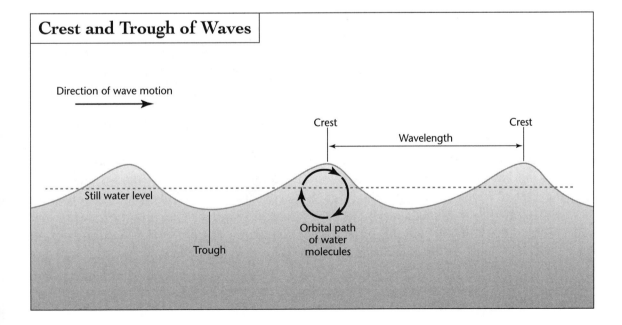

Tides

Tides result from a combination of three forces: the gravitational force of the Sun, the gravitational force of the Moon, and the motion of the Earth. Gravity is the force of attraction, or pull, between two bodies. Everything that has mass exerts gravity. The Earth and Moon exert gravitational pulls on each other. Because the Earth has more mass than the Moon, its gravity keeps the Moon in orbit. The Moon does not fall into the Earth because of the inertia, the tendency of a moving object to keep moving, that is created by their stable orbits.

The inward force of gravity and the outward force of inertia affect the entire surface of the Earth, but not to the same degree. Owing to Earth's rounded shape, the equator is closer to the Moon than Earth's poles are. The pull of the Moon's gravity is consequently stronger around the equator. On the side of the Earth facing the Moon at any given time, the Moon's gravity pulls the Earth toward it. The solid Earth is unable to respond dramatically to that pull, but the liquid part of Earth can. As a result, the ocean bulges out toward the Moon on the side of Earth that is facing it. On the side that is farthest from the Moon, inertia flings water away from the Moon. The Moon's pull on one side of Earth and the force of inertia on the opposite side create two bulges—high tides—in the ocean.

The bulges do not rotate around the Earth as it turns on its axis. Instead, they remain aligned with the Moon as the Earth rotates under them. Different parts of the Earth move into and out of these bulges as it goes through one rotation, or one day.

Even though the Sun is much farther from Earth than the Moon is, the Sun also has an effect on tides. The Sun's influence is only about half that of the Moon's. A small solar bulge on Earth follows the Sun throughout the day, and the side of the Earth opposite the Sun experiences a small inertial bulge.

The Moon revolves around the Earth in a 28-day cycle. As it does so, the positions of the Moon, Earth, and Sun relative to one another change. The three bodies are perfectly aligned during two phases: new moon and full moon, as shown in Figure 1.4. At these times, the Sun and Moon forces are acting on the same area of Earth at the same time, causing high tide to be at its highest and low tide to be at its lowest. These extremes are known as spring tides and occur every two weeks.

During first- and third-quarter conditions, when only one-half of the Moon is visible in the night sky, the Sun and Moon are at right angles to the Earth. In these positions, their gravitational pulls are working against each other, and the two bodies cancel each other's effects to some degree, causing high tides to be at their lowest, and low tides to be at their highest. These neap tides also occur every two weeks.

Physical Aspects 17

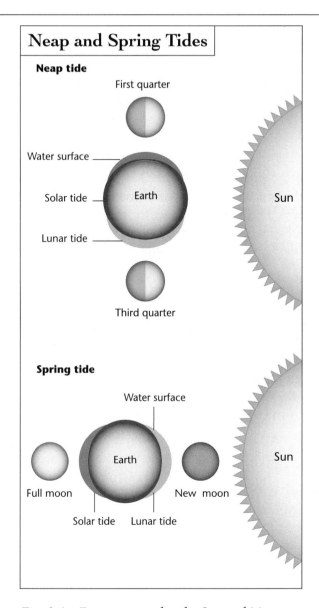

Fig. 1.4 *Every two weeks, the Sun and Moon are aligned with Earth so that the gravitational forces of both heavenly bodies create very high tides called spring tides. When the Moon and Sun are at right angles to Earth, lower, or neap, tides result.*

landslides, volcanic eruptions, and movement along faults on the ocean floor.

As shown in Figure 1.3, the path followed by water particles in a wave is called an orbit. Water depth affects a wave's behavior because it acts on the orbit. A wave traveling through deep water has a circular orbit. Near shore, the orbit strikes the bottom and its shape becomes distorted. As the round orbit shortens, the wave's energy gets packed into less water depth. At the same time, friction created when the lower portion of the wave rubs the bottom slows it. Eventually, the top or crest of the wave moves ahead of its supporting base, and the wave tilts forward and then falls over, or breaks. After the wave, or surf, breaks, the water rushes to shore. The surf zone is the region between the point where waves start breaking and their furthest reach.

Other causes of waves include undersea landslides, volcanic eruptions, and movement along faults on the ocean floor. Fault movements can suddenly raise or lower the seabed, resulting in earthquakes and displacing water. The displaced water forms a gigantic wave called a tsunami (sometimes referred to as a seismic or tidal sea wave).

Tsunamis are similar in structure to the waves generated by wind but differ dramatically in size. At the most, a wind driven wave has a wavelength, the distance from the top of one wave to the top of the next wave, of a few hundred yards. The wavelength of a tsunami can be more than 100 miles (160.9 km) long.

At the surface of the open sea, a tsunami is not noticeable because it is spread over such a large area. As the wave passes over the sea, it may displace the water only a few feet. Since wave crests created by a tsunami are widely separated, ships at sea rarely notice the traveling water. From an airplane or in satellite photographs, such a wave is virtually invisible.

Tsunamis are most evident along coasts. The leading edges of tsunamis can form walls of water thousands of miles long. As water arrives at the coast, it slows, causing the waves to compress and direct their energy upward. When the tall waves break on the shore, the walls of water crash against the coast with destructive force.

On December 26, 2004, the most powerful earthquake in the past 40 years, registering 9.0 on the Richter scale, struck near the northern tip of Sumatra. Tsunamis generated by the quake radiated around the Indian Ocean. Within minutes, Sumatra was pounded by waves more than 32.8 feet (10 m) tall. Somalia, on the opposite side of the ocean basin, was blasted with 13.1-foot (4-m)-tall waves eight hours later.

The tsunamis were devastating, killing more than 200,000 people. In addition to the extensive loss of human life and property, coastal ecosystems were ravaged. After being ripped apart by tons of water, the shores were inundated with sediment and debris. Depending on the extent of the damage, recovery time for these ecosystems will range from a few months to a century.

Even though tsunamis are dramatic, they are rare. Most of the water in the ocean is moved in wide, marine rivers called currents. Winds create currents that move near the water's surface, so many currents follow the same paths as wind belts. There are dozens of ocean currents, all named for their average positions. The Gulf Stream flows northward along the eastern coast of North America, carrying warm tropical water with it.

The California Current flows southward along the western coast, moving cold, North Pacific water toward the equator.

Tides are the periodic and predictable rising and falling of large bodies of water. The entire ocean is affected by tides, but their impacts are most obvious in coastal areas. These regular changes in water level create environments along coasts that are alternately submerged in water or exposed to air.

Coastal Habitats

Habitats along coasts are subjected to a variety of conditions. Some areas are always underwater, some are flooded once or twice a day, and others are rarely flooded. The fact that seawater rushes into and away from the coast creates an abundance of unique habitats there.

The zone of the coast between the highest and lowest tides is called the intertidal, or littoral, zone. This space is the only section of the marine environment that is ever exposed to air. As a result, organisms that live in the littoral zone deal with problems that other marine organisms never face, such as regular periods of desiccation during each low tide. To avoid being left high and dry, animals that are capable of locomotion travel back and forth with the ebbing water. Less mobile organisms must stay in one place, but they possess techniques and structures to prevent drying. Some are covered in waterproof shells or mucous secretions that conserve moisture. Many of the shelled animals possess flaps called opercula that seal the organisms inside their shells when the tide is out. An animal without an operculum may secrete a layer of mucus to seal its shell. There are almost as many adaptations for conserving moisture as there are kinds of animals along the coast.

In the littoral zone, temperatures can range to extremes. Some organisms have specialized adaptations that help moderate fluctuations in temperature. Many animals, for example, live in light-colored shells that reflect heat rather than absorb it. Others have sculptured shells with extensions that help dissipate heat.

Besides dryness and extreme temperatures, littoral zone residents are exposed to a variety of salinities. In shallow water, a rain shower can add enough freshwater to the system to dramatically reduce salinity. On the other extreme, a period of dryness increases the evaporation of water, leaving the shallows in a saltier-than-usual condition.

Perhaps one of the most difficult factors of littoral life is the constant pounding of waves. Life-forms have developed adaptations that help them survive the smashing, throwing, and beating action of water as it breaks against the shore. Some hold to rocks or plants so they will not wash away. Many have smooth, streamlined bodies that offer little resistance to the action of the water. Other animals cement their bodies to firm surfaces, securing themselves in place even during the worst wave action.

Conclusion

Coastlines are formed by geologic processes over long periods of time. When a coast is young, all of the processes that affect it are the ones associated with land. Rains dissolve minerals in rocks and soil, glaciers dig troughs and push moraine to the coast, rivers carry soil to the sea, and sections of the Earth's crust crack, leaving gaps. As a coast ages, it is also impacted by natural forces in the ocean, such as waves and water. Coasts are both built and eroded by waves and currents, and organisms such as corals and mangrove trees make further changes.

In some places, coastlines are rocky, varying in structure from cliffs to piles of boulders. In other areas, they are sandy beaches or wide mud flats. The residents of these ocean edges are exposed to extremely stressful living conditions that vary in salinity, levels of dissolved gases, and temperature.

Salinity, the level of dissolved minerals in water, has a tremendous impact on the kinds of organisms that can live in a coastal habitat because it affects how materials move into and out of cells. Most coastal organisms are exposed to varying salinity due to input or removal of water. Levels of dissolved gases in the intertidal water can also vary. Organisms that require oxygen can be stressed if levels of the gas drop radically. The temperature of shallow coastal water can fluctuate widely over just a few hours. Temperature has an impact on how quickly, or slowly, biological processes proceed in organisms. To survive these harsh and unpredictable conditions, organisms that live in the coastal environment are highly specialized.

Today's coastlines are the results of millions of years of sculpting by physical, chemical, and geological changes. These conditions have yielded some of the most treasured areas on Earth and produced a vast and unimaginable array of adaptations in organisms that live there.

2

Microbes and Plants
Bacteria, Protists, Plants, and Fungi Along the Coast

A quick glance at a shoreline might reveal only foamy surf ebbing over sand, rocks, or mud. The areas of ocean between the highest high tides and the lowest low tides often seem uninhabited. Yet, these ecosystems are some of the most productive environments in the sea, sustaining a diversity of living things that interact through complex food webs.

The health and vitality of every ecosystem is directly related to its supply of nutrients, and the seacoast is rich in this regard. More nutrients can be found along the shore than in any other part of the marine environment. Nutrients support the living things there and form the bases of all food chains. On the coast, nutrients arrive from two sources. One source is the group of resident producers, all of the organisms that are green and able to carry out photosynthesis. These provide food for plant-eating animals. The other source is the organic material that washes or blows into the area from both land and sea. Organic material, like rotting plant and animal tissue, is consumed by bacteria and fungi.

The kinds of substrates in intertidal zones determine, to some degree, the types of organisms that can live there. Generally, intertidal substrates are sandy, rocky, muddy, or a combination of these types. Along each type of substrate, the transition of physical factors between shore and sea is not uniform. As a result, living things form four distinct zones, or bands, that are largely dictated by how often water covers the area.

The highest intertidal area is the splash zone, a region that is only submerged by tides during storms. Generally, the only seawater this zone receives is moisture from the splash of waves. Serving as a transition from land to water, this region is rarely submerged, so the organisms that live there are more adapted for life on land than in the sea. The splash zone is

Food Chains and Photosynthesis

Living things must have energy to survive. In an ecosystem, the path that energy takes as it moves from one organism to another is called a food chain. The Sun is the major source of energy for most food chains. Organisms that can capture the Sun's energy are called producers, or autotrophs, because they are able to produce food molecules. Living things that cannot capture energy must eat food and are referred to as consumers, or heterotrophs. Heterotrophs that eat plants are herbivores, and those that eat animals are carnivores. Organisms that eat plants and animals are described as omnivores.

When living things die, another group of organisms in the food chain—the decomposers, or detritivores—uses the energy tied up in the lifeless bodies. Detritivores break down dead or decaying matter, returning the nutrients to the environment. Nutrients in ecosystems are constantly recycled through interlocking food chains called food webs. Energy, on the other hand, cannot be recycled. It is eventually lost to the system in the form of heat.

Autotrophs can capture the Sun's energy because they contain the green pigment chlorophyll. During photosynthesis, detailed in Figure 2.1, autotrophs use the Sun's energy to rearrange the carbon atoms from carbon dioxide gas to form glucose molecules. Glucose is the primary food or energy source for living things. The hydrogen and oxygen atoms needed to form glucose come from molecules of water. Producers give off the extra oxygen atoms that are generated during photosynthesis as oxygen gas.

Autotrophs usually make more glucose than they need, so they store some for later use. Heterotrophs consume this stored glucose to support their own life processes. In the long run, it is an ecosystem's productivity that determines the types and numbers of organisms that can live there.

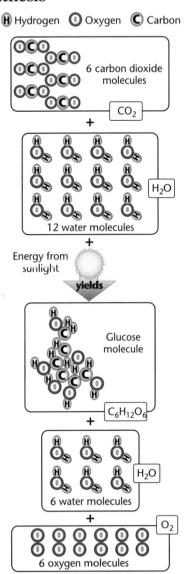

Fig. 2.1 *During photosynthesis, the energy of sunlight is used to rearrange the components of carbon dioxide and water molecules to form glucose, water, and oxygen.*

usually bare of plants, although some grasses and lichens may grow there.

Moving toward the ocean, the next area, the high intertidal zone, extends from the lower end of the splash zone down to the point where extreme high tides reach. Here, the organisms are most often uncovered, although they may endure periods of flooding. Living things in this region face the most extreme conditions. The few organisms that make their homes in the high intertidal zone are hardy and are more land adapted than sea adapted.

Continuing toward the sea, the next region is the middle intertidal zone, the broadest and most variable. Life-forms in the middle intertidal zone are routinely covered and uncovered with water. Each time water flows out, they are exposed to hours of drying conditions caused by sunlight and wind. Although this is a tough place to live, a wide variety of organisms make their home here, and most of them are equipped with adaptations for both land and water.

Closest to the sea is the low intertidal zone, where living things are submerged most of the time. Since these organisms are rarely exposed to drying conditions, they are completely adapted for life in the sea. Of all parts of the intertidal region, this zone has the greatest number of species.

Monerans

The most numerous organisms living on the seashore are monerans. This group consists of one-celled organisms, both autotrophs and heterotrophs that are extremely simple in structure. All of the monerans play essential roles in shoreline ecosystems. The autotrophic forms are some of the most important intertidal producers, and the heterotrophs are vital to the decomposition processes.

All of the autotrophic monerans are collectively known as cyanobacteria. Because they contain chlorophyll, they can use the Sun's energy to make their own food. In addition to chlorophyll, these organisms possess two other pigments: phycocyanin, a blue pigment, and phycoerthrin, a red one. The presence of these two accessory pigments helps cyanobacteria capture more of the Sun's wavelengths than green chlorophyll could do alone.

As one-celled organisms, individual cyanobacteria are inconspicuous. Some species form colonies and filaments that look like slimes, velvety coatings, mats, or tufts. Their colors range from green to brown, black, blue green, and red. Hundreds of species of cyanobacteria live along the coast. Some of the more common genera (groups of species) are *Lyngbya, Phormidium,* and *Calothrix.*

Cyanobacteria are dominant producers on some shores, especially in the temperate and tropic zones, where they live among grains of sand and soil. Several species of the single-celled organisms are capable of moving with a simple gliding action, so they migrate toward the surface of the sand when light conditions are low. When the sun is hot, they burrow into the sand to prevent drying. Usually these cells are invisible to the naked eye, but they can form large clumps or mats on the shore.

Calothrix is a genus of cyanobacteria that produces what is known as the "black zone," a line of black, tarlike blobs, in the high intertidal zone. Each globule is a colony of cyanobacterial cells. These colonies can endure extended periods without water because they are contained in gel-like sheaths that keep them moist. *Calothrix* also grows like a crust on rocks and on other types of algae in many intertidal regions.

A different cyanobacteria, *Lyngbya aestuarii,* forms smooth mats along the shore. Individual cells secrete thick sheaths that give the algae a wet, jellylike quality.

Ancient Cyanobacteria

The oldest-known photosynthetic organisms, the cyanobacteria may form long, mucus-producing filaments that trap and hold debris and grains of sand. In the earliest stages of life on Earth, sticky strands of algae, with their ensnared soil particles, created gray towers, or stromatolites, that rose several yards upward from the seafloor. In the past, millions of stromatolites flourished near coastlines. Today, only a few of these primitive structures are still alive, and they are found in very salty, warm-water habitats.

Cyanobacteria are believed to be the ancestors of plant cells. In the ancient seas, heterotrophic cells probably fed on cyanobacteria, engulfing them whole. Instead of digesting these prey, the heterotrophic cells were able to maintain them in membrane-bound sacs. The cyanobacteria continued to photosynthesize, making their own food and providing food for their new hosts. Over time, these two individual cells became an inseparable team, dependent on each other for support. The cyanobacteria acted as the light-gathering structure for the two, and the duo evolved into primitive plant cells.

Smooth mats of *Lyngbya* can be found where intertidal flats are regularly covered with water. Within a mat, the cells are so tightly packed that oxygen cannot diffuse into it. As a result, anaerobic bacteria develop colonies below the mats that appear as purple or black layers.

Heterotrophic Bacteria

Not all bacteria contain chlorophyll. Those that do not, the heterotrophic bacteria, must ingest their food. Heterotrophic bacteria feed on all kinds of organic matter, including dead or decaying plants and animals, other bacteria, and dissolved nutrients. As simple single-celled organisms, heterotrophic bacteria do not have mouths for taking in food. To feed, they release digestive enzymes on their meals. After the enzymes

Kingdoms of Living Things

There are millions of different kinds of living things on Earth. To study them, scientists called taxonomists classify organisms by their characteristics. The first taxonomist was Carolus Linnaeus (1707–78), a Swedish naturalist who separated all creatures into two extremely large groups, or kingdoms: Plantae (plants) and Animalia (animals). By the middle of the 19th century, these two kingdoms had been joined by the newly designated Protista, the microscopic organisms, and Fungi. When microscopes advanced to the point that taxonomists could differentiate the characteristics of microorganisms, Protista was divided to include the kingdom Monera. By 1969, a five-kingdom classification system made up of Monera (bacteria), Protista (protozoans), Fungi, Animalia, and Plantae was established. The five-kingdom system is still in use today, although most scientists prefer to separate monerans into two groups, the kingdom Archaebacteria and the kingdom Eubacteria.

Monerans are the smallest creatures on Earth, and their cells are much simpler than the cells of other living things. Monerans that cannot make their own food are known as bacteria and include organisms such as *Escherichia coli* and *Bacillus anthracis*. Photosynthetic monerans are collectively called cyanobacteria, and include *Anabaena affinis* and *Leptolyngbya fragilis*. In the six-kingdom classification system, the most common monerans, those that live in water, soil, and on other living things, are placed in the kingdom Eubacteria. Archae-

break the food down into simple molecules, bacterial cells absorb them.

As a group, heterotrophic bacteria are vitally important to the process of decomposition. Along the coastline, hundreds of different kinds of bacteria break down organic matter that washes into the region. Heterotrophic bacteria live in all areas of the intertidal zone, including in the shallow water, atop and within the sediment, and on plants. By breaking down complex compounds in the intertidal ecosystem, they provide food for themselves, and they release inorganic nutrients such as phosphorus and nitrogen back into the environment. In addition, bacteria serve as a source of food for organisms like protozoans, small crustaceans, and mussels.

bacteria are the inhabitants of extreme situations, such as hot underwater geothermal vents or extremely salty lakebeds.

Another kingdom of one-celled organisms, Protista, includes amoeba, euglena, and diatoms. Unlike monerans, protists are large, complex cells that are structurally like the cells of multicellular organisms. Members of the Protista kingdom are a diverse group varying in mobility, size, shape, and feeding strategies. A number are autotrophs, some heterotrophs, and others are mixotrophs, organisms that can make their own food and eat other organisms, depending on the conditions dictated by their environment.

The Fungi kingdom consists primarily of multicelled organisms, like molds and mildews, but there are a few one-celled members, such as the yeasts. Fungi cannot move around, and they are unable to make their own food because they do not contain chlorophyll. They are heterotrophs that feed by secreting digestive enzymes on organic material, then absorbing that material into their bodies.

The other two kingdoms, Plantae and Animalia, are also composed of multicelled organisms. Plants, including seaweeds, trees, and dandelions, do not move around but get their food by converting the Sun's energy into simple carbon compounds. Therefore, plants are autotrophs. Animals, on the other hand, cannot make their own food. These organisms are heterotrophs, and they include fish, whales, and humans, all of which must actively seek the food they eat.

On some sandy beaches, enormous colonies of bacteria create large splotches of color. These microbial mats, called colored sand, or Farbstreifensandwatt, are multilayered. The first layer is made up of dense filaments of cyanobacteria. Beneath the cyanobacteria, heterotrophic sulfur bacteria such as *Thiocapsa roseopersicina* thrive. Sulfur bacteria are organisms that get the energy they need for cellular reactions from sulfur compounds instead of from the Sun. In the process, they produce by-products such as smelly hydrogen sulfide gas or iron compounds that stain the sand black. Sulfur bacteria are anaerobic organisms, meaning that they cannot survive in the presence of oxygen, whereas most of the organisms that grow on top of the sulfur bacteria layer are described as aerobic, because they must have oxygen to survive.

In the summer, *T. roseopersicina* may undergo rapid growth and create large populations, or blooms, in intertidal zones. These can appear along with blooms of *Lyngbya aestuarii*. The thick, gelatinous covering created by *L. aestuarii* holds in moisture and makes the mass dense, preventing oxygen from diffusing into the sand below it. As a result, it creates a perfect environment for *T. roseopersicina* residing beneath it.

Protists and Fungi

Protists are larger, more complex single-celled organisms than bacteria. Many of the autotrophic protists are key producers in the intertidal environments while some of the heterotrophic forms are important consumers. Diatoms and ciliates are two common types of coastal protists.

Diatoms are one-celled organisms that contain chlorophyll, plus a brown-gold light-absorbing pigment that has earned them the nickname "golden algae." Some forms are free floating, while others attach to the substrate or plants. There are hundreds of types of diatoms, but they share many common characteristics. Most types build a protective box around themselves called a frustule.

The frustules of diatoms are made of many minerals, but the primary component is silicon, which is also found in sand and glass. Each microscopic box is a delicate, transparent cage

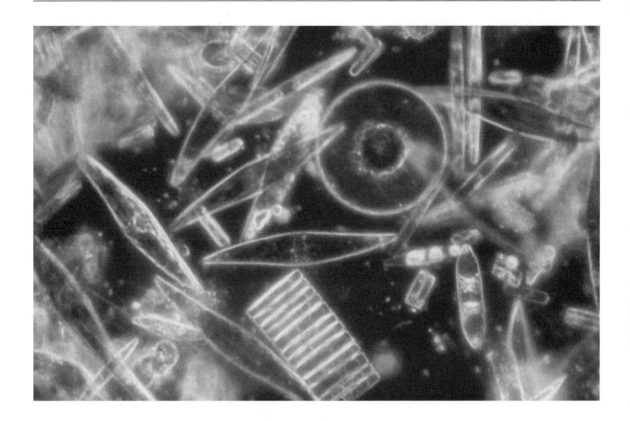

Fig. 2.2 *Diatoms are single-celled organisms that may be round or elongated.* (Courtesy of Dr. Neil Sullivan, University of Southern California, NOAA)

that protects the cell inside but lets in enough light to permit photosynthesis. Frustules are perforated with hundreds of tiny openings that allow the organism within to interact with the watery environment. As Figure 2.2 shows, the basic shape of a frustule can be either elongated or round, depending on the species. Details of frustules, like the patterns created by the perforations, vary by species. Most of the elongated forms are found in the sand and sediment, while the round ones are more common in the water column.

As in many protists, reproduction in diatoms is by binary fission, an asexual method in which the parent cell divides into two identical cells. Asexual reproduction involves only one parent, and the two offspring are clones, or exact duplicates of the parent. During fission, a diatom cell experiences challenges that cells of other species of protists do not face; it also has to divide its frustule.

When a parent diatom cell splits apart to form two "daughter" cells, each daughter inherits one-half of the parent's frustule. The inherited halves become lids for each daughter cell, and both daughters grow a new lower half. As a result, one cell is the same size as the parent, and one is smaller. Over several generations, cells and frustules become too small to divide further. To solve the problem, the organisms leave their old, undersized shells, and one of two things happens. The cells either undergo a period of growth and then secrete new, spacious frustules or go through sexual reproduction.

Sexual reproduction can occur in several ways. In some species, a small diatom cell breaks into little pieces, each of which swims around until it finds another diatom cell with which it can fuse. The product of their fusion builds the new frustule. In other species, two adult diatom cells line up beside each other. Each cell undergoes division, and then they exchange one daughter cell. The new pairs of daughter cells fuse, resulting in two new cells, each possessing genetic material from two parents.

Advantages of Sexual Reproduction

Even though asexual reproduction seems like a simple solution to continuing a species, many monerans and protists also undergo sexual reproduction. While asexual reproduction expands a population, it does not make it possible for the population to change in any way. All of the organisms created in asexual reproduction are clones, so they have the same genetic information and the same characteristics as the parent organism. As long as environmental conditions remain steady, asexual reproduction maintains a healthy population; however, if anything in the environment changes, the population may suddenly be at risk. Because all the individuals are alike, any problem that may befall one cell will probably visit them all, possibly resulting in the loss of the entire population.

In organisms that reproduce sexually, all of the offspring are different. Each one contains a unique set of genetic information, half of it inherited from one parent and half from the other. Since individuals in the population vary, it is unlikely that a change in the environment would create problems for everyone. In fact, a change that reduces the survival rate of some might improve the survival rate of others.

Populations of intertidal zone diatoms are enormous, and they can be found in every imaginable sunlit habitat. Diatoms grow attached to other plants or on the bodies of animals. In addition, some of the coastal species flourish on and in the sand and sediment or attached to surfaces and cavities of rocks.

Diatoms living in the surf of the littoral zone are constantly on the move, migrating back and forth with the water. Offshore, they travel on air bubbles from the bottom of the shallow water to its surface. As a wave rolls to shore, diatoms ride it in and wash up on the beach. Here they secrete sticky mucus that helps them sink just below the sand's surface. As a wave rushes back to ocean, the diatoms wash out with it, back to their offshore position. From here, they float to the surface and reenter on another wave.

Instead of migrating to and from shore or hiding in the spaces between sand grains, some diatoms spend their entire lives floating in the shallow water. To maintain positions near the water's surface and in the light, they have developed several strategies to stay afloat. Tiny sacs of oil within the cells keep some species afloat since oil is less dense than water and will float above it. Other species have very lightweight frustules that float easily. Many types of diatoms are equipped with projections and wings that slow their rate of sinking.

Along both temperate and tropical shores, an unusual variety of diatom forms mucuslike sheaths, or tubes, that hold colonies of cells. These tubes resemble the long, thin threads of filamentous brown algae, but a close look under a microscope shows that the tubes hold large diatom cells. Unlike free-living diatoms, the cells of most tube-dwelling species are not covered with silicon frustules.

Another type of protist commonly found along the intertidal zone is the ciliate, a one-celled organism covered with short, hairlike projections called cilia that are used for movement and for feeding. Usually solitary and free-swimming, these organisms feed on other protists and bacteria and are eaten by very small animals, such as worms. As predators of bacteria, ciliates are vitally important in returning nutrients to the intertidal ecosystem. Bacteria take in enormous amounts of organic matter floating or beached in the intertidal zone.

When ciliates consume the bacteria, the nutrition stored in the organic matter becomes available to the rest of the food chain. Ciliates also excrete nitrogen and phosphorus compounds, making these inorganic materials available to the bacterial population and helping to perpetuate the food web.

One ciliate, *Strombidium capitatum,* is a heterotroph that feeds on single-celled organisms with chloroplasts. Instead of digesting the chloroplasts of its prey, *S. capitatum* keeps them alive and functional. In this way, this particular ciliate has two systems for getting food: to prey on other organisms or to consume the products of photosynthesis.

Fungi, like bacterial decomposers, break down organic matter that accumulates in the intertidal zones. As they grow on the tissues of dead or decaying organisms, fungi send out tiny filaments called hyphae. Each filament releases enzymes that dissolve the tissues, creating a digested liquid that fungal cells can absorb.

Plants of the Coast

Plants are abundant along many shorelines because some intertidal zones provide excellent growing conditions. Inorganic nutrients that accumulate from the decomposition of organic matter serve as food, and the plants receive plenty of light in the shallow water zones.

Conversely, the intertidal environment creates unique challenges to plants. Some parts of the coast are regularly pounded by waves and exposed to strong winds. Plants that can live in these environments are highly adapted for them, with strong attachments to the soil and blades that are limber and not easily damaged by moving water. More plants grow in low-energy areas of the coast than in the surf zones.

Marine plants fall into two major types: macroalgae and vascular plants. Macroalgae, or seaweeds as they are commonly called, are multicellular autotrophs with unique adaptations for their marine habitats. Seaweeds occur in three different colors: green, red, and brown, depending on the pigments they contain. Many types of macroscopic algae are found in the intertidal zone, whereas vascular plants are rare there and are primarily represented by a few species of marine grasses.

Light and Algal Coloration

Light is a form of energy that travels in waves. When the Sun's light arrives at Earth, it has a white quality to it. White light is made up of the colors red, orange, yellow, green, blue, indigo, and violet. The color of light is dependent on the length of the light wave. Light in the visible spectrum contains colors and has wavelengths between 0.4 and 0.8 microns (1 micron equals $\frac{1}{1,000,000}$ of a meter, or .000001 m; a micron is also known as a micrometer). Violet light has the shortest wavelength in the visible spectrum and red has the longest.

Light is affected differently by water than it is by air. Air transmits light, but water can transmit, absorb, and reflect light. Water's ability to transmit light makes it possible for photosynthesis to take place beneath the surface. All of the wavelengths of visible light are not transmitted equally, however; some penetrate to greater depths than others.

Light on the red side of the spectrum is quickly absorbed by water as heat, so red only penetrates to 49.2 feet (15 m). Blue light is not absorbed as much, so it penetrates the deepest, reaching 100 feet (33 m). Green light, in the middle of the spectrum, reaches intermediate depths. When light enters water that is filled with particles such as dirt and plant matter, as in an estuary, it takes on a greenish brown hue because it only penetrates far enough to strike, and be reflected from, the particles. In tropical water where particulate levels are very low, light travels much deeper before it reaches enough particles to be reflected back to the surface, so tropical water appears blue. Below 1,500 feet (457.2 m), no light is able to penetrate.

Because of the way light behaves in water, aquatic plants do not receive as much of the Sun's energy as do plants on land. To compensate, most species contain some accessory pigments, chemicals that are adept at capturing blue and green light. These accessory pigments provide the plants additional light and thereby help macroalgae increase their rate of photosynthesis. Some of these pigments mask the green of chlorophyll and give colors to macroalgae that are not usually associated with plants. Accessory pigments explain why seaweed occurs in shades of brown, gold, and red. Green algae contain accessory pigments, too, but they do not mask the color of chlorophyll as the pigments in other kinds of algae do.

Green Algae

Two common genera of green macroalgae are *Ulva* and *Enteromorpha*. *Ulva*, which grows near the low tide mark, resembles bright green, translucent sheets of tissue paper with ruffled edges. Depending on the species, the length of

Differences in Terrestrial and Aquatic Plants

Even though plants that live in water look dramatically different from terrestrial plants, the two groups have a lot in common. Both types of plants capture the Sun's energy and use it to make food from raw materials. In each case, the raw materials required include carbon dioxide, water, and minerals. The differences in these two types of plants are adaptations to their specific environments.

Land plants are highly specialized for their lifestyles. They get their nutrients from two sources: soil and air. It is the job of roots to absorb water and minerals from the soil, as well as hold the plant in place. Essential materials are transported to cells in leaves by a system of tubes called vascular tissue. Leaves are in charge of taking in carbon dioxide gas from the atmosphere for photosynthesis. Once photosynthesis is complete, a second set of vascular tissue carries the food made by the leaves to the rest of the plant. Land plants are also equipped with woody stems and branches that hold them upright so that they can receive plenty of light.

Marine plants, called macroalgae or seaweeds, get their nutrients, water, and dissolved gases from seawater. Since water surrounds the entire marine plant, these dissolved nutrients simply diffuse into each cell. For this reason, marine plants do not have

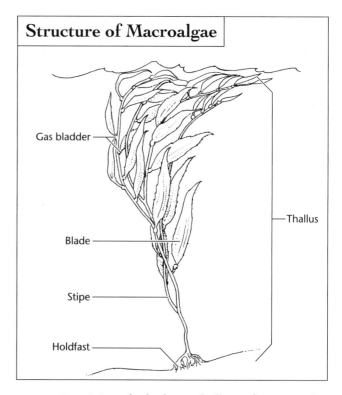

Fig. 2.3 The body, or thallus, of a macroalga is made up of leaflike blades, stemlike stipes, and rootlike holdfasts. Gas bladders on the stipes and blades help hold the plant near the top of the water column.

vascular tissue to accommodate photosynthesis or to carry its products to each cell. In addition, marine plants do not need support structures because they are held up by the buoyant force of the water. Since water in the ocean is always moving, the bodies of marine plants are flexible, permitting them to go with that movement. Some marine plants secrete mucus to make their surfaces slick, further reducing their drag or resistance to water movement. Mucus also helps keep animals from eating them.

A plant that grows on land is described with terms such as *leaf, stem,* and *root*. Seaweeds are made up of different components, which are shown in Figure 2.3. The parts of seaweed that look like leaves are termed *blades,* or fronds. Some are equipped with small, gas-filled sacs, or *bladders,* that help keep them afloat and close to the sunlight. The gases in these bladders are usually nitrogen, argon, and oxygen. The stemlike structures of macroalgae are referred to as *stipes*. A root-shaped mass, the *holdfast,* anchors seaweeds but does not absorb nutrients like true roots do. Together, the blades, stipes, and holdfast make up the body, or *thallus,* of the macroalgae. Thalli take on many different forms, including tall and branched or thin and flat.

sheets varies from 11.8 inches (30 cm) to 39.4 inches (1 m). These extremely thin sheets of alga are made up of only two cell layers. A small, disc-shaped holdfast secures most species of *Ulva* to the substrate, but free-floating species also exist.

Reproduction in *Ulva* is triggered by lunar cycles. Like most algae, *Ulva* reproductive strategy involves an alternation of generations: One generation yields gametes, and the next produces spores. Male and female reproductive parts are located on separate plants. During the gametophyte phase, specialized cells along the edges of blades, leaflike structures, convert to reproductive cells that release sperm and eggs into the seawater. The gametes combine to form zygotes, cells with complete sets of DNA that grow into new sporophyte plants. DNA is the genetic material in a cell that controls its activities. When they mature, sporophyte plants develop reproductive cells that undergo meiosis, a type of cell division that reduces the genetic material within the cell by one-half. These cells form spores that grow into new male and female gametophyte plants, and the cycle begins anew.

Ulva can be found on all the world's coasts. Known commonly as sea lettuce or green laver, the alga usually grows in dense colonies. Many species are used as ingredients in soups and salads, and as a substitute for nori, the popular seaweed in sushi.

Another intertidal green alga is *Enteromorpha intestinalis*. The bright green tubular fronds of this alga are filled with gas, then constricted at irregular intervals, giving them the appearance of intestines. The fronds grow 3.9 to 11.8 inches (10 to 30 cm) long from small holdfasts that are secured to the soil or rocks. Gases contained in the fronds help them float on the surface of the water.

E. intestinalis will grow on any bare spot in the tidal area and is very tolerant of a wide range of physical conditions. Reproduction is accomplished by forming spores on the edges of blades. The spores are released during a rising tide, ensuring that offspring will be widely dispersed. This alga is a summer seaweed that dies back in cold weather, leaving masses of white fronds on the beach.

Several species of *Codium* grow as thick, spongy dark green tubes that resemble fingers and are known by names such as dead man's fingers, felty fingers, and sea staghorn. *Codium fragile,* common on rocks in the middle to low intertidal zones along coasts around the world, has a dark green color and forms Y-shaped, cylindrical branches that are 0.2 inch (0.6 cm) in diameter and 11.8 inches (30 cm) tall. The plant is secured in place by a wide, spongy holdfast. The entire *C. fragile* plant is made up of one very long cell. This unusual type of cell, called a coenocyte, has multiple nuclei but no cell walls separating them. Like other green algae, *Codium* reproduces by forming gametophyte and sporophyte generations. They can also regenerate asexually from fragments.

Brown Algae

Brown algae are found in most parts of the ocean, including the intertidal zones. The colors of these plants, which range from yellow to black, are due to combinations of chlorophyll and accessory pigments. Most types of brown algae produce chemicals that protect them from predators.

Several species of *Fucus,* a group of brown algae commonly known as rockweed, are found in the intertidal zone. Brown, slippery fronds of *Fucus* grow to 20 inches (50 cm) in length. *Fucus* generates toxins that make the seaweed difficult to digest and unattractive to many grazers. The plant does not maintain a supply of toxins in its fronds. Instead, the protective chemicals are manufactured after grazing begins, an adaptation that allows some grazing but prevents predators from destroying the plant. For this reason, snails and small crustaceans munch a little on each plant without destroying any of them. This strategy of producing toxins as needed may conserve the alga's energy resources, since toxin production is a high-energy task.

Fucus is highly adapted for its unstable, shallow water environment. Fronds are equipped with air bladders that keep the plant afloat when the tide comes in. The stipes and fronds synthesize a gel-like material, alginate, which improves flexibility, allowing the plant to move with the flow of water. Alginate also conserves moisture, preventing desiccation when the tide is out.

Knotted wrack (*Ascophyllum nodosum*) and bladder wrack (*Fucus vesiculosus*) are common coastal brown algae. Bladder wrack can be distinguished by a midrib in the frond and pairs of air bladders. Reproductive structures are wartlike knobs on the ends of the fronds. Knotted wrack lacks a midrib but has the same kinds of paired air bladders and knobby reproductive structures. In both cases, gametes are released into seawater where they unite.

Hanging wrack (*Bifurca brassicaeformis*) is a close relative that creeps along the ground and often covers rocks. Spear-shaped blades produce eggs and sperm along their margins. Eggs hang from the edges of blades by delicate mucus threads that prevent them from being washed out of the surf zone and into an area where they might not germinate.

Fucus provides an important refuge for small animals that live in the intertidal zone. When the tide goes out, marine animals need a place to hide and stay moist until the water returns. Many spend the low tide hours hiding underneath *Fucus* fronds. The oases provided by these large blades create short-lived communities of organisms that crowd together temporarily, very much like a group of pedestrians who gather under a canopy to wait for a rainstorm to pass. When the tide returns, the coastal organisms hiding beneath *Fucus* scatter.

Plant Defenses

Grazers like fish, crabs, and snails feed on macroalgae. Even though the sea plants do not have the option of running or hiding when attacked, they are not defenseless victims. Many of these plants have evolved proactive mechanisms for protecting themselves. Some produce noxious chemicals that either taste bad or make the grazers sick. Usually after only a bite or two, animals feeding on chemically defended seaweeds will leave them alone and look for something else to eat. Predators learn very quickly which seaweeds are safe and tasty and which are not.

Other species of macroalgae protect themselves by depositing calcium carbonate in their outermost tissues. Calcium carbonate gives tissue a hard, rocklike covering that keeps a lot of predators from nibbling the seaweeds. Although most animals ignore calcified macroalgae, a few predators are able to bite through the hard layer. These animals may nevertheless avoid eating the seaweeds because their stomachs cannot process the chemicals, which cause production of uncomfortable carbon dioxide gas.

Many species of macroalgae use a combination of physical and chemical defenses. *Halimeda* deposits calcium carbonate in its cell walls, and it also produces chemicals. In this case, the chemicals are not created until after a predator takes a few bites, a strategy that conserves energy.

Other brown algae that grow along the coast include the sea cauliflower (*Leathesia difformis*), which lives atop the fronds of other plants, and the sea palm (*Postelsia palmaeformis*), a stalklike plant that resembles a palm tree. Sea palms can grow on top of mussels, increasing the likelihood that the small animals will be uprooted from their positions by wave action. When this happens, the vacancy on the substrate created by a mussel's departure is quickly filled by another sea palm.

Red Algae

There are more than 4,000 species of red algae, so called because they contain accessory light-catching pigments in various shades of red. Red algae range from simple one-celled organisms to large multicellular structures. In the intertidal area, many species are branched, a strategy that exposes the maximum amount of surface area to water.

Purple laver (*Porphyra capensis*) is a red alga found high on the shore as purple, membranous sheets. Depending on the species, these plants vary from 4 to 20 inches (10 to 50 cm) in length. The reproductive cycles of laver are very complex and include a two-year alternation of generations. In the gametophyte stage, pink-edged female plants and yellow-edged male plants release gametes into the water. During another phase of their life cycle, *Porphyra* plants exist as microscopic organisms inside the shells of bivalves, such as mussels and barnacles.

One group of red algae forms a hard, skeletal-like covering of calcium carbonate. Known as coralline algae, these reds are the most prevalent of all the calcium-depositing marine plants. One member of this group, *Lithothamnion*, looks like a pink, stony crust growing over rocks in tide pools from New England to the Florida Keys. *Corallina* is another type of calcified red alga that grows erect instead of encrusting on rocks. The branches of *Corallina* are flexible because they are not calcified between the segments.

Irish moss (*Chondrus crispus*) is a small, purplish-colored red seaweed that grows up to 7.8 inches (20 cm). Underwater, Irish moss fronds look iridescent, but in very strong light, they may take on a green color. The plant is commercially important for the gel-like carbohydrate in its cell walls, which is used to make food additives, textile sizing, and cosmetics. In Japan, the same chemical is used in the manufacture of shampoo because it gives a gloss to hair. Consequently, Irish moss is often harvested from lower intertidal zones.

Sea sacs (*Halosaccion glandiforme*) are red algae that develop saclike fronds. Growing up to 3 feet (0.9 m) in length, sea sacs are easy to spot in the intertidal zone. Their long structures fill with water during high tide and lose water during low tide. Other common intertidal red algae include the *Bossiella* species, the branching coralline algae that are pink to purple, and *Mazzaella spendens*, an iridescent purple to green seaweed with flat blades.

Sea Grasses

Along with the green, brown, and red macroalgae, another type of plant grows in the intertidal zone: several species of

Fig. 2.4 Kelp and surfgrass grow in the tidal zone. (Courtesy of Dan Howard, Sanctuary Collection, NOAA)

grasses. These true plants first evolved to live on the land, then moved back to the sea, so they have many of the typical terrestrial plant adaptations like roots, vascular tissue, pollen, and seeds. Once a year, these underwater grasses bloom with thousands of tiny, pollen-producing flowers. After pollination, fertilized eggs mature into seeds that float away from the parent plants, sink, and start new beds of grass.

Surfgrass (*Phyllospadix*) grows up to 3 feet (0.9 m) tall in rocky and low intertidal areas (see Figure 2.4). Although a true plant, surfgrass has some unique adaptations that help it survive in the intertidal zone. Unlike other vascular plants, surfgrass lacks stomata, tiny openings in the leaves through which water vapor enters and exits the leaves. The absence of stomata helps this grass retain water. The outer surface of the plant is thickened to protect it from damage by energetic moving water. Because this particular grass is very high in protein, it is an important food for many of the animals in the intertidal zone.

Conclusion

The seashore is a harsh environment where inhabitants are exposed to extremes in temperature and salinity, as well as alternating periods of drying and flooding. During a period of just a few hours, conditions such as the water level can fluctuate widely. The temperature in a small pool of intertidal water can vary as much as 86°F (30°C) or suffer a 50 percent change in salinity.

Despite these problems, the intertidal zone is rich in nutrients and able to support more living things than most other parts of the ocean. Every nook and cranny, even the spaces between the sand grains, brim with communities of organisms.

Key players along the coast are the producers, primarily diatoms, seaweeds, and grasses. Diatoms, microscopic organisms in tiny glasslike shells, live in the water column or on the sediment where they carry out photosynthesis. Macroscopic algae, or seaweeds, can be green, brown, or red, depending on their combination of pigments. All of the intertidal algae are highly adapted to the constant flux of the environment. Some are protected with a calcified skeleton, while others produce alginate, a compound that makes them slippery and flexible enough to move with the water.

Heterotrophic bacteria and fungi are critical to the intertidal zone because they break down the dead or decaying matter that accumulates there. The activities of heterotrophic bacteria return nutrients to the system, making it possible for the producers to thrive. Fungi and heterotrophic bacteria are also important members of the food chain, sustaining other protists as well as small animals such as worms and copepods.

Rocks along the edges of the sea are coated with slick layers of yellow-brown seaweed that are secured in place with holdfasts. Fronds of decomposing seaweed left by the receding water are boons to marine organisms that are stranded during low tide, providing shade, moisture, and food to a variety of creatures. The interdependence of different types of organisms on the coast is typical of food webs throughout the marine environment.

3

Sponges, Cnidarians, and Worms
Simple Invertebrates in Coastal Waters

Invertebrates are the most numerous animals on Earth. All members of this large group share one common characteristic: They do not have a backbone. In the coastal habitats of the intertidal zones, invertebrates such as sponges, corals, anemones, jellyfish, worms, clams, mussels, snails, and limpets make up the majority of the animals. Some are very primitive creatures whose bodies lack the simplest forms of organization. Others are more advanced, with organs and body systems that are highly sophisticated.

The intertidal zone is a perfect home for many species of invertebrates. Because the area is rich in nutrients, oxygen, and light, it provides plenty of food for small animals who are filter feeders as well as those who are predators. The challenges of these environments require that the invertebrates who reside there be specialized. Pounding waves, periods of little or no water, extreme salinities, and varying temperatures are stressful environmental conditions for all inhabitants.

Among the simplest invertebrates are the sponges, cnidarians, and worms. Sponges are the most primitive of the three groups and represent the transition between colonies of cells and animals with organized tissues. Cnidarians are more complex than sponges, equipped with mouths and simple digestive cavities. Worms, the most developed of the three types of organisms, have definite head and tail ends, specialized body parts, and simple organ systems.

Sponges

Looking more like plants than animals, sponges are primitive organisms that lack tissues, such as skin and bone, as well as organs, such as a brain or heart. Instead, individual cells within sponges carry out all body functions (see Figure 3.1).

Sponges, Cnidarians, and Worms 43

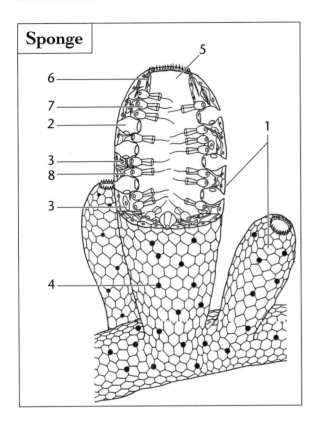

Fig. 3.1 The epidermis (1) of a sponge is filled with tiny pores called porocytes (2). Amoebocytes (3) move around the sponge carrying food to cells. Water enters the sponge through an incurrent pore (4), flows into the central cavity, and exits through the osculum (5). Spicules (6) lend support to the sponge's body wall. Choanocytes (7) lash their flagella in the central cavity to keep water moving through the sponge and to gather bits of food that are suspended in the water. The mesoglea (8) is a jellylike matrix located between the epidermis and the cells that line the central cavity.

Anatomically, sponges are made of only two layers of cells, an outer epidermis and an inner gastrodermis. Sandwiched between these two layers is a jellylike layer, the mesoglea.

Sponges contain structures that support the cell layers and give the animals shape. In many types of sponges, these structures are small, mineralized needles called spicules that are scattered throughout the mesoglea. Instead of spicules, some species possess fibers made of a tough, rubbery protein called spongin.

Many of the body functions that are performed by tissues and organs in higher animals are carried out by amoebalike cells in sponges. When food is digested, these cells pick up the nutrient molecules and share them with other cells by traveling throughout the body of the sponge, crawling through the mesoglea and both cell layers. The cells also perform jobs such as collecting and removing waste products as well as arranging spicules in their proper places.

Sponges eat by gathering tiny food particles that are suspended in the water, then processing them to extract the nutrients. The gastrodermis of a sponge is covered with a layer of collar cells, or choanocytes, that are equipped with long, whiplike flagella. By lashing their flagella back and forth, chaonocytes create currents of water that flow in through tiny pores in the epidermis, move through the gastrovascular cavity, and flow out one or more openings called oscula. As water travels through the gastrovascular cavity, chanocytes capture tiny food particles that are suspended in it.

Sponges occur in a variety of forms. In the intertidal zone, their shapes are often dictated by the conditions in which they live. Sponges in high-energy water, where the wave action is strong, may be thin and flat. Those in quieter water may grow into vase or finger shapes.

As adults, sponges are sessile, spending all of their time in one place. Also, adult sponges are hermaphrodites, organisms that produce both male and female sex cells. During reproduction, specialized sex cells form eggs and sperm. Sperm are discharged into the water and travel to another sponge where they fertilize its eggs. In some species, both eggs and sperm are released, and they unite in the water instead of inside the sponge. The zygotes that result from fertilization become free-swimming larvae that eventually settle to the bottom, attach to the substrate, and grow into adult sponges. A sponge can also reproduce asexually by forming buds, small organisms that grow at the base of the parent sponge. When buds are mature, they separate from the parent and live independently. In addition, if a sponge is torn apart by waves, each part can develop into a new organism.

Sponges attract many species of hungry animals but are not easy prey. Their needlelike spicules offer them some protection against predators. In addition, most species produce chemicals that discourage other animals from eating them.

The toxins also keep plant spores or animal larvae from growing on their surfaces.

One sponge species found in many coastal zones is the breadcrumb sponge (*Halichondria panicea*). Green to yellow crusts of the sponge creep across rocks and can cover an area of several feet. Breadcrumb sponges can be found as thin mats in the energetic surf zones, but in quiet tide pools, they form thicker masses.

Purple sponge (*Haliclona permollis*), a gray to pale violet encrusting form, grows on rocks and algae. Sometimes purple sponge develops interconnected, raised branches that spread to 4 inches (10 cm) and stand 1 inch tall (25 mm).

Oplitaspongia pennata is a thin, reddish-colored encrusting sponge that reaches 36 inches (91 cm) in width. The surface of this species has a smooth texture that is dotted with tiny, star-shaped pores. *O. pennata* prefers areas where there is strong wave action, so it grows well on tide pool rocks. A red nudibranch, a sluglike organism whose color matches the color of the sponge, is one of its primary predators.

Organ-pipe sponges (*Leucosolenia*) create vaselike white to yellow colonies that grow to be 0.35 inch (9 mm) tall and 0.06 inch (1.6 mm) in diameter. Another erect, although much larger, species is the finger sponge (*Haliclona oculata*), which reaches heights of 18 inches (45 cm), with branches that are 0.5 inch (1.2 cm) in diameter. The finger sponge, which has numerous slender branches, can occur in a variety of colors, including yellow, brown, and purple. Pieces of the sponge commonly break off and wash up on the shore.

Yellow patches on the shells of mollusks are boring sponges (*Cliona*), organisms that drill holes and craters into shells, especially those of oysters. If the oyster shells contain living inhabitants, these sponges may cause their death. Often, boring sponges completely overgrow and dissolve empty shells and then are forced to live independently. Although individual animals are 0.5 inch wide, boring sponges may form colonies that can grow to 12 inches (30 cm) in diameter.

Body Symmetry

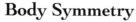

An important characteristic of the body plan of an animal is its symmetry. Symmetry refers to the equivalence in size and shape of sections of an animal's body. Most animals exhibit body symmetry, but a few species of sponges are asymmetrical. If a plane were passed through the body of an asymmetrical sponge, slicing it in two, the parts would not be the same.

Some animals are radially symmetrical. Shaped like either short or long cylinders, these stationary or slow-moving organisms have distinct top and bottom surfaces but lack fronts and backs, heads or tails. A plane could pass through a radially symmetrical animal in several places to create two identical halves. Starfish, jellyfish, sea cucumbers, sea lilies, and sand dollars are a few examples of radially symmetrical animals.

The bodies of most animals are bilaterally symmetrical, a form in which a plane could pass through the animal only in one place to divide it into two equal parts. The two halves of a bilaterally symmetrical animal are mirror images of each other. Bilateral symmetry is associated with animals that move around. The leading part of a bilaterally symmetrical animal's body contains sense organs such as eyes and nose. Fish, whales, birds, snakes, and humans are all bilaterally symmetrical.

Scientists have special terms to describe the body of a bilaterally symmetrical animal, depicted in Figure 3.2. The head or front region is called the anterior portion and the opposite end, the hind region, is the posterior. The stomach or underside is the ventral side, and opposite that is the back, or dorsal, side. Structures located on the side of an animal are described as lateral.

Fig. 3.2 A sponge (a) is an asymmetrical animal. Starfish and jellyfish (b) are radially symmetrical; snails, turtles, and fish (c) are bilaterally symmetrical. In a bilaterally symmetrical animal, the head or front end is described as anterior and the tail end as posterior. The front or stomach side is ventral and the back or top side is dorsal. The sides of the animal are described as lateral.

Sponges, Cnidarians, and Worms 47

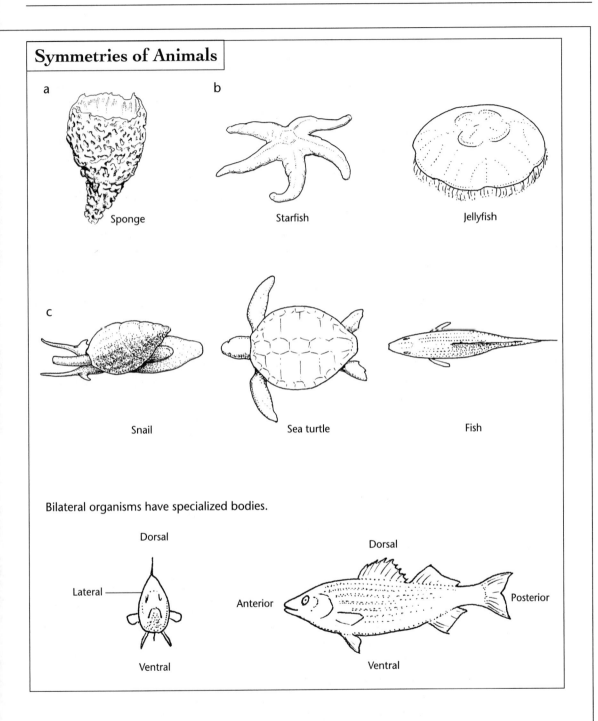

Cnidarians

Cnidarians are a group of animals that includes hydrozoids, sea fans, anemones, and jellyfish. All of the organisms in this group have simple, saclike bodies in the polyp or medusa forms (see Figure 3.3). The bell-shaped bodies of jellyfish are referred to as medusae, and the vaselike structures that are typical of coral and anemones are polyps.

The cnidarian body has only one opening, the mouth, through which food is taken in and wastes are expelled. The mouths of cnidarians are surrounded with rings of tentacles. The body wall is made of two layers, the epidermis and endodermis, with a jellylike mesoglea between them.

A cnidarian is equipped with a simple network of nerves that runs through its body. Specialized cells called cnidocytes, located in the tentacles, tie into the network of nerves and play roles in defense and in the capturing of food. Each cnidocyte is armed with a nematocyst, a barb attached to a long filament. When triggered, the nematocyst uncoils and shoots out its barbed tip. In some species, the tips contain poisons that can paralyze or kill prey, and in others, they are covered with sticky mucus. Once a victim is snared, tentacles move it through the cnidarian's mouth and into the gastrovascular cavity. There, enzymes break down the prey, and cells in the gut cavity absorb the nutrients. Undigested parts are expelled through the mouth.

Anemones are cnidarians that develop as polyps. The body of an anemone is a thick column with two distinctive ends, as shown in the lower color insert on page C-1. The upper end, or oral disk, is a ring of tentacles around a narrow slit at the center, the mouth. Grooves beside the mouth continuously bring in water to provide oxygen to interior tissues. Depending on the species, oral disks vary in diameter from a fraction of an inch to 12 inches (30 cm).

The other end of the animal, the pedal disk, attaches to hard substrates with a suction-cup-like action. Some secrete a sticky adhesive to help hold them in place. Although the animals are classified as sessile, or sedentary, they actually can move around very slowly, either by shuffling across intertidal spaces on their pedal disks, or by somersaulting on their tentacles from place to place. If food is scarce, anemones may pump their bodies full of water and float to a new location, looking very much like jellyfish while in transit.

Anemones are well adapted to their high-energy habitats. Their soft, flexible bodies easily absorb the impact of waves. Most of the animals attach themselves to one place and remain there throughout life, traveling little, if at all. Many intertidal species form colonies, an adaptation that helps them retain water when the tide goes out. At low tide, the animals draw in their tentacles to conserve moisture and don "hats" of light-colored shells and stones to keep cool and slow down the rate of evaporation. These "hats" reflect heat and light, trap moisture, and serve as camouflage to protect anemones from predators.

Many species of anemones live in a close association with a dissimilar organism. These symbiotic relationships are usually with simple green algae or dinoflagellates. Dinoflagellates are protists that inhabit the tissues of several types of simple animals. By living together, both the host animals and the

Fig. 3.3 Cnidarians have two body plans: either a vase-shaped polyp (a) or a bell-shaped medusa (b). Each plan is equipped with tentacles (1), a gastrovascular cavity (2), and a single body opening, the mouth (3).

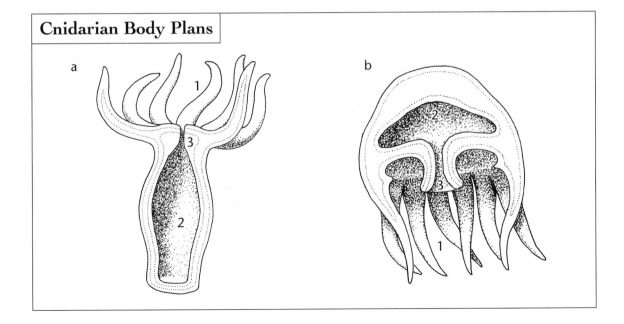

Cnidarian Body Plans

dinoflagellates benefit. The protists supply food and oxygen for their hosts, while the anemones protect the dinoflagellates from predators and provide a place to live that gets plenty of exposure to sunlight. The relationship between dinoflagellates and anemones is not unique; there are other organisms that also support microscopic producers. As a group, symbiotic one-celled autotrophs that live in the tissues of plants or animals are known as zooxanthellae.

The aggregate green anemone (*Anthopleura elegantissima*) is very common in rocky intertidal areas. Easily recognized by its green body, pink-tipped tentacles, and radiating lines on the oral disk, an aggregate green anemone may occur as a solitary individual or as a member of a large group, referred to as an aggregate. The solitary form usually grows to 6 inches (15 cm) in diameter, while the aggregate individuals are smaller, only 0.8 to 2 inches (2 to 5 cm) wide. For protection, the animal is frequently covered with bits of shell and stone.

The aggregate green anemone hosts both green algae and dinoflagellates as symbiotic residents, but like all anemones, it can also catch its own food with its tentacles. Prey, which include a number of small animals, such as copepods, isopods, and amphipods, are either stung by nematocysts or ensnared in sticky mucus on the tentacles. Once the prey is subdued, the anemone pushes it through the mouth and into the gastrovascular cavity.

A. elegantissima is an example of an anemone that can reproduce sexually and asexually. In the sexual phase, both eggs and sperm are released during the summer months, usually in July. Eggs are fertilized in the gastric cavity, and then the young are released through the mouth as free-swimming larvae. After a short time, the larvae settle to the bottom to grow into adult anemones. Asexual reproduction occurs by longitudinal fission, a process in which the organism splits down the middle. The two resulting individuals crawl away from each other. Fission can yield aggregates of anemones that are pressed closely together, often with several hundred individuals.

The giant green anemone can grow to almost 1 foot (30 cm) in height and 9 inches (25 cm) in diameter. The column of this colossal species varies from green to brown, and its tentacles are green, blue, or white. Found in tide pools and channels of water that form on exposed rock, this anemone may form groups, with individuals close enough to touch tentacles. Muscles, crabs, urchins, and small fish are among its preferred prey.

The sea anemone known as *Tealia coriacea* has a brown or purple oral disk surrounded by tentacles that taper to a point and are tipped in magenta. The column may be orange or red, and the lower portion is often mottled. Usually, only the tentacles are visible because these animals attach themselves to buried objects, such as rocks or shells, so that they live partially submerged in the sediment. When threatened, *Tealia coriacea* retract their entire bodies into the sediment.

The small, bright red strawberry anemone (*Corynactis californica*), shown in Figure 3.4, forms colonies in areas where the substrate is covered with rocks. The anemone's tentacles end in knobs that are pink and transparent. Although the strawberry anemone can reproduce sexually, it often divides by fission to form large colonies that cover 3 feet (0.9 m) or more of sand.

Fig. 3.4 The strawberry anemone waves its tentacles in the water. (Courtesy of Getty Images)

The thenarian burrowing anemones (*Actinothoe*) are a group of long, wormlike cnidarians that lie buried within the soil. Each animal is equipped with numerous tentacles around its oral disk. Under the oral disks, the anemone's column extends below the surface of the soil to attach to a stone or pebble.

Another type of cnidarian that lives in the intertidal zone is the hydroid, an animal that differs from anemones in two ways: Hydrozoids exist in both the polyp and medusa forms, and they form colonies in which individual polyps have specialized duties, dividing the labor needed to support the group. In colonies, some individuals assume responsibility for reproduction while others take charge of tasks such as

Spawning and Brooding

For sexual reproduction to take place, male and female cells must come together. Many marine species spawn, or discharge one or both of their sex cells into the water. For this strategy to be successful, eggs and sperm must be released at the same time, which is why spawning usually occurs once a year at a specific time. Animals are cued to release gametes by specific environmental factors, such as the Moon's phase, length of daylight, or temperature.

The alternative strategy to spawning is brooding. Animals that brood release sperm into the water, while eggs remain within the mother. Sperm swim around until they find a female, enter her body, and fertilize the eggs. Eggs are brooded within the mother's body until time for them to hatch.

Fertilized eggs that are brooded have the advantage of protection from predators during development. In comparison, eggs fertilized in the water column or on the seafloor are at high risk from predators. For this reason, animals that brood their developing eggs only produce small numbers of gametes, while those that spawn discharge hundreds of thousands of gametes, a strategy that ensures that a few of the resulting offspring will survive until adulthood.

feeding or defense. All of the polyps in a hydrozoan colony are connected with a shared gastrovascular cavity. Colonies are usually very small and can be found attached to such surfaces as rocks. The individual animals within a hydrozoan colony are tiny, measuring about 0.04 inch (1 mm) in width. Many species support autotrophic zooxanthellae in their tissues. Like anemones, hydroids can also capture prey with stinging tentacles.

Hydroids have very complex reproductive strategies. Special structures on polyps produce offspring in the medusa stage. Medusae then release eggs and sperm. Fertilized eggs develop into swimming larvae, which settle down to begin new hydrozoan colonies.

Two common groups of hydroids are tubularians and pennarians. Tubularian hydroids (*Tubularia*) create colonies of dense, pinkish growths on most solid intertidal substrate, including pilings, buoys, and jetties. Organisms in the colonies form unbranched stems that hold flowerlike polyps. In colonies of *Tubularia indivisa*, individual polyps can reach heights of 4 to 6 inches (10 to 15 cm). The plantlike colonies are made up of straight, erect stems that are joined at the base, each stem covered in a tough, yellow material. The polyps inside are pink or red and crowned with 50 to 100 tentacles.

Feather hydroids (*Pennaria tiarella*) form branching colonies that grow to 6 inches (15 cm). The polyps are tipped with knobby tentacles set in four or five whorls around the mouth. Feather hydroids can be found growing on eelgrass and wooden pilings in the intertidal zone.

Unlike hydrozoids, jellyfish are cnidarians that spend their entire lives in the medusa form. Most live in water deeper than the intertidal area, but individuals sometimes stray or wash into the shallow water of the surf. One very unusual type of jellyfish is native to the intertidal zone, the stalked jellyfish (*Haliclystus*). Instead of the usual medusa form, the body of the stalked jellyfish is attached to the substrate by a stalk on which the medusa is flipped upside down, with the mouth facing up, giving the animal the appearance of a flower. The "petals" of the flower are eight lobes that are tipped with clublike tentacles. *Haliclystus auricula* is one kind of stalked jellyfish whose lobes end in hundreds of tentacles. This very small jellyfish, reaching $1\frac{3}{4}$ inches (4.45 cm) in height, favors quiet tide pools where it may be found attached to snails, seaweed, or rocks. *H. auricula* feeds on worms and other small shore animals.

Another group of cnidarians, the octocorals, form colonies similar to those of hydrozoids. The individual polyps in these colonies resemble anemones, but the colonial animals live together inside a strong, protective matrix. One species, dead man's fingers (*Alcyonium digitatum*), varies in color from white to pink to orange and is characterized by fleshy lobes and fingerlike projections. The 4- to 8-inch (10 to 20 cm) colonies attach to stone and shells. Another octocoral, the sea whip (*Leptogorgia virgulata*) forms colonies of slender, whiplike branches that can be purple, tan, or red.

Worms

Marine worms show a remarkable degree of adaptability and have evolved to occupy many intertidal niches. Some species are free swimming, but most live quiet, inconspicuous lives in the sand and beneath rocks or shells. Worms feed in a variety of ways, grazing on algae, preying on other animals, and scavenging dead and decaying material.

The populations of worms living in intertidal areas fall into two major groups: flatworms and segmented worms. Of the two types, flatworms are the simplest and most primitive. A typical marine flatworm is a tissue-thin, unsegmented animal that glides across the substrate on hairlike cilia. A simple, highly branched digestive tract shows through the animal's translucent body. The digestive system is designed with only one opening, so food enters and undigested matter exits through the same place. To feed, a flatworm extends its pharynx, a muscular tube, onto its meal. The tube secretes digestive enzymes on the food, creating a soupy mix of partially digested tissue that is sucked into the intestine. Most species are carnivores that prey on small invertebrates and protozoans.

Flatworms may reproduce in several ways. Some divide asexually by fission, producing two identical offspring. Other species have all-female populations whose eggs develop without fertilization. However, most types of flatworms reproduce sexually. Individual flatworms are hermaphroditic, having both male and female reproductive organs. To cross-fertilize their gametes, two worms copulate, each donating sperm to the other.

Marine flatworms generally have very poor eyesight. Their "eyes" are actually sets of light-sensitive eyespots, arranged in pairs of two, four, or six on the anterior end of the body. To get information about the environments, the worms also depend on folds of tissue that create tentacle-like structures. These tissues can detect chemicals in the water, helping the worms navigate and find their food.

The oyster flatworm (*Stylochus ellipticus*) lives in tide pools and under rocks. This cream-colored organism has a narrow band of eyespots and a fringe of tiny tentacles on its anterior end. Oyster flatworms, which can grow up to 1 inch (2.5 cm) long, feed on barnacles as well as oysters.

The speckled flatworm (*Notoplana atomata*) lacks tentacles but has eyespots arranged in four clusters. On the dorsal side, the worm is brown with small flecks of color. Like the oyster flatworm, the speckled flatworm is small, usually less than 1 inch (2.5 cm).

The giant of the flatworm world is the milky ribbon worm (*Cerebratulus lacteus*), which grows to 3 to 4 feet (0.9 to 1.2 m) long and 0.5 inch (12 mm) wide. Most of the year, the animal is milky colored, but during breeding season, the males become bright red and females change to brown. This worm has sensory grooves located on its head and a very small mouth on the underside. During the day, the animal hides under rocks, emerging at night to move around on a trail of slime.

Although flatworms are common in the intertidal zone, populations of segmented worms, or polychaetes, greatly outnumber them. All segmented worms have bristles, or setae, along their external surfaces. Setae usually occur in bundles as part of the structure of the polychaete foot, or parapodium. In some worms, parapodia are tiny bumps, but in others they are lobed appendages with gills and fingerlike projections. Generally, parapodia are more developed in swimming worms than in burrowing ones.

Sexes are separate in polychaetes. During reproduction, sperm and egg are released in water where they unite to form zygotes. Zygotes grow into planktonic larvae that swim in the shallow water for a short time. Eventually, they settle to the bottom and metamorphose into the adult forms.

Some of the polychaetes in the intertidal zone are described as free swimming, or errant. These organisms spend much of their time wandering around the intertidal zone actively looking for food. The body of an errant species is long and slender and looks about the same from head to tail, although the head end of the animal usually has specialized structures for feeding and respiration, including a muscular proboscis—a tubular feeding and sucking organ—and jaws. Typically, errant organisms are carnivores that spend much of their time foraging for food but rest in tubes that they build in the sediment. Examples of errant species include the sea mouse, many-scaled worm, twelve-scaled worm, bloodworm, clam worm, and paddle worm.

The sea mouse (*Aphrodita hastata*) is a short worm with a furlike, iridescent coat made of modified setae. Beneath the coat, there are 15 pairs of scales. This animal reaches lengths of 6 inches (15 cm) and may be 3 inches (7.5 cm) wide.

The many-scaled worm (*Lepidasthenia commensalis*) is a slender, 4-inch (10-cm) animal that is covered with 30 to 50

Worm Comparisons

Segmented worms are much more advanced and complex than flatworms. The digestive systems of flatworms are one-way tubes sandwiched between two body walls. However, segmented worms have a space between their two body walls called the body cavity, or coelom, that represents an important evolutionary advance, one that provides a place for the body's internal organs. In segmented worms, organs are held in their proper places inside the coelom by a membrane, the peritoneum.

All animals with coeloms are equipped with one set of muscles around the body wall and another set around the digestive system. The body wall muscles help the animal move about, while the digestive system muscles push food along the digestive tract. In contrast, flatworms have only one set of muscles in their body wall, and these muscles must carry out both functions.

Segmentation is an advance in animal evolution because segmented animals can increase in size by adding more body portions. In addition, segments can become specialized to carry out certain jobs. Flatworms are therefore limited in size as well as in the degree of specialization they can reach because they lack segments.

A flatworm gets oxygen and loses carbon dioxide by simple diffusion through the epidermis. Segmented worms have more complex gas exchange systems. Oxygen diffuses through the skin into blood vessels. Blood then carries oxygen to cells deep in the

---

pairs of scales. Close relatives, the twelve-scaled worms (*Lepidonotus*) have only 12 to 13 pairs of scales. Colors vary from brown to yellow, red, or green depending on the species. When disturbed, these organisms roll up into a ball. Some live inside the shells of flat-clawed hermit crabs and among oyster shells, sharing space with their residents.

A bloodworm (*Glycera*), shown in the upper color insert on page C-2, looks red because its body fluid shows through the worm's pale skin. The tapered head holds four antennae and a large proboscis armed with four small black fangs that can be withdrawn. Some species grow to 15 inches (19.5 cm) in length. Bloodworms are abundant on tidal flats, where they can use their toothed proboscis to quickly bury themselves into the ground. The worms can also bite, causing an injury similar to a bee sting.

worm's body, while it picks up carbon dioxide and carries it out of the body. The blood of segmented worms contains hemoglobin, an iron-containing compound that attracts oxygen and binds to it. Blood with hemoglobin is capable of carrying 50 times more oxygen than blood that lacks the molecule. Near the head of the segmented worm, five pairs of muscular vessels or hearts squeeze rhythmically to keep blood circulating through the worm's body.

Segmented worms have much more advanced digestive systems than flatworms do. A flatworm has one opening, a mouth, for food and wastes. A segmented worm has two openings, a mouth at one end and an anus at the other. The mouth opens to an esophagus that leads to a muscular pharynx. Food travels from the pharynx to the crop where it can be stored temporarily before entering the gizzard, an organ that grinds it. From there, food goes to the intestine, the site of digestion. Digested nutrients enter the bloodstream, and waste materials are expelled through the anus. Segmented worms also have special organs called nephridia that remove nitrogen wastes from blood and excrete those wastes through tiny openings in the body wall.

The evolutionary advancements from flatworms to segmented worms are reflected in other animals such as mollusks and crustaceans, as well as in vertebrates. The segmented worms, although still evolutionarily simple, provided the groundwork from which further advancement evolved.

The syllid worms (which belong to the genus *Syllis* or the genus *Autolytus*) have two distinct stages of life: breeding and nonbreeding. During the nonbreeding stage, they are active predators that creep along the bottom looking for food. Measuring about 2 inches (5 cm) long, these worms have three antennae and two pairs of tentacle-like extensions from the head. During the breeding stages, usually during summer months, they grow epitokes, reproductive extensions that are filled with sperm or eggs, depending on the gender. A new head forms at one end of the epitoke, then the entire segment breaks off as a free-swimming worm. In the water column, the skin binding the epitoke dissolves, freeing the gametes. Thousands of worms release their epitokes on the same evening, filling the water with worm gametes. Eggs and sperms unite in the water to form zygotes.

Clam worms (*Nereis*) have well-developed parapodia and four or more pairs of tentacle-like extensions on their heads. During the day, they hide in mucus-lined sand burrows. At night, they emerge to search for prey, usually small crustaceans or mollusks. The worms will also scavenge any dead animals they can find.

Others polychaetes are considered to be sedentary because they stay in one place, waiting for food to come to them. Their bodies are often shorter than errant worms, with clearly defined regions. Like the sabellid worm in the lower color insert on page C-2, sedentary polychaetes spend their lives in tubes or holes and have developed a variety of special adaptations for feeding and breathing in these niches. Most of these worms consume plankton or other small material that is suspended in the water. Some species of sedentary polychaetes include the coiled tube worm, sand builder worm, trumpet worm, parchment worm, lugworm, and hard tube worm.

Sand builder worms (*Sabellaria*) are sedentary segmented worms that construct hard calcium tubes in which to live. Groups of worms create reeflike structures that may measure 1 foot (30 cm) or more in length. A collar near the worm's head secretes the material that hardens to form a tube. As the worm grows, its collar continually enlarges one end of its long home. The tubes of sand builder worms are coiled, giving them the appearance of small, fat snails. Like all sedentary polychaetes, the worms have developed specialized structures for feeding and breathing. Gases are exchanged on tentacles that the worms extend from their tubes. Hairlike cilia on the tentacles create currents that draw in plankton and other small food items, which the tentacles trap and transport to the mouth. During low tide, these worms stay moist by plugging the openings to their tubes with tentacles that are modified to be opercula.

The trumpet worm (*Pectinaria gouldii*) has a head that is flattened in the front and surrounded by rings of iridescent golden setae. Measuring only about 0.6 inch (1.52 cm), this small worm constructs curved, cone-shaped tubes from grains that it cements together. Trumpet worms dig head first into the sand, eating sediment as they progress. Organic matter in the sediment is digested, and the dirt is defecated.

Named for the parchment-like material from which it builds its tube, the parchment worm (*Chaetopterus variopedatus*) lives in a U-shaped home dug in mud, both ends of which are marked by piles of dirt. The body of a parchment worm is a flabby structure that is divided into three regions. Three paddle-like parapodia in the middle section create a current of water and plankton through the tube. A netlike bag of mucus hangs near the head end of the worm and filters plankton from the water. When the bag is full of food particles, a special organ transfers it to the worm's mouth along a ciliated groove. The mucus bag method of feeding allows the parchment worm to very efficiently remove all of the food in the water it processes.

A lugworm (*Arenicola*) is a stout animal whose front half is much thicker than its back half. The 8-inch (20-cm) long body is divided into three regions. The small head and tail lack appendages, but the trunk, which starts about one-third of the way back, possesses bundles of setae and a tuft of gills on each side. Like parchment worms, lugworms dig U-shaped burrows with openings at both ends. A worm ingests sand with its proboscis and then backs out of its burrow to discharge the processed dirt into a pile of casting by the door of its home. Lugworms also consume particles of food brought in by water.

The bamboo worm (*Clymenella*) is aptly named, as it looks very much like a 4-inch (10-cm) length of bamboo. The body is made up of a small number of long segments with very few parapodia. Bamboo worms build sand- or mud-encrusted tubes and consume mud to obtain nutrients.

Conclusion

The populations of invertebrates found in intertidal waters are diverse and include hundreds of species. Invertebrates, animals that lack backbones, are generally small and occur in a wide variety of body designs. Some of the less complex invertebrates include sponges, cnidarians, and worms.

At one time, the plantlike, sessile lifestyles of sponges left scientists wondering whether these organisms were plants or animals. Now, sponges are known to be extremely simple animals whose bodies consist of a colony of cells. The body wall, a loose matrix of mesoglea between two layers of cells, is supported by spicules or spongin. Currents created by the beating flagella of chaonocytes draw water into the sponge through pores in the body wall. Food particles stick to the chaonocytes, which digest them and pass the nutrients on to amoebalike cells to be shared with the rest of the organism. Reproduction can be sexual or asexual by budding or fragmentation.

Cnidarians represent a higher level of complexity than sponges. Their bodies are saclike structures with only one opening to the digestive system. Tentacles around the mouth are used for food gathering as well as for defense. Anemones are common in tide pools, where many form colonies that are efficient at conserving water and dispersing the force of the waves.

Several species of stalked, plantlike hydrozoans and octocorals form small colonies in shallow intertidal waters. These organisms spend part of their lives as medusae and part as polyps. Like anemones, they gather food with their small tentacles. A less common but more conspicuous cnidarian along the shore is the jellyfish. Most jellyfish found in the intertidal zone are simply washed in, although the stalked jellyfish makes its home there.

Worms are plentiful in the shallow intertidal waters. Both simple flatworms and more complex segmented worms can be found there. Because of its size, the milky flatworm may be the most obvious, but the shoreline also supports many other types, including the speckled flatworm and oyster flatworm.

Segmented worms are generally found along the bottom, living in deposits of sediment or under rocks. Some types build tubular homes for themselves in which they spend their entire lives. These sedentary worms include the parchment tube worm and the trumpet tube worm. Errant species, such as bloodworms and syllid worms, spend some of their time out of their hiding places, actively seeking prey.

All of the simple invertebrates that live in the shallow waters of the coast are highly adapted for the stressful living conditions there. Many, for example sponges and hydrozoans, are considered to be sessile because they spend all of their adult lives in one place. Others, such as the worms and anemones, are motile, moving from one place to another. Feeding styles in the intertidal zones include filter feeding, scavenging, and hunting. To supplement their diets, some animals have long-term symbiotic relationships with one-celled green organisms that are capable of making food. In all cases, simple invertebrates in the intertidal zones are a highly successful and varied group of animals.

4

Mollusks, Arthropods, and Echinoderms
Complex Coastal Animals

Invertebrates make up the largest classification of organisms on the land as well as in the ocean. To more easily understand such a sizable group, these animals can be divided into two subgroups: simple invertebrates and complex invertebrates. The division is somewhat arbitrary, being loosely based on the body plans of the animals involved. On the coast, simple marine invertebrates—sponges, anemones, and worms—share space and food with their more sophisticated relatives, which include, clams, whelks, horseshoe crabs, shrimp, true crabs, starfish, and sea cucumbers.

A very simple organism, such as a sponge, does not require a body system to get oxygen to its cells. A sponge is able to absorb all the oxygen it needs directly from the environment. However, more complex animals are too large to meet all of their oxygen needs in this way. In marine environments sophisticated respiratory systems include gills.

Gills are respiratory organs made up of thin tissues that are densely packed with tiny blood vessels. The tissues of gills are tightly folded, packing a large surface area into a small space. As water flows over gills, oxygen that is dissolved in it diffuses into the blood in the capillaries. At the same time, carbon dioxide dissolved in the blood diffuses into the water and is carried out of the body.

A complex invertebrate also has a circulatory system to carry dissolved oxygen to and remove carbon dioxide from each of the millions of cells that make up its body. To feed these cells, the animal has a digestive system that takes food into the body and breaks it down, then turns the nutrients over to the circulatory system for delivery. To keep blood flowing to all of the cells on a full-time basis, a muscular heart powers the circulatory system. An excretory system takes care of removing wastes from cells. All of these body systems are run and coordinated by the brain, a complex arrangement of nerve cells on the anterior end of the animal.

Mollusks

Mollusks are complex invertebrates whose name literally means "soft bodied." Even though all mollusks possess soft bodies, their outward appearances vary enormously, from small animals enclosed in two hard shells to large, eight-legged creatures that hide quietly among the rocks. Mollusks are subdivided into four groups based on their anatomy: chitons, which are insectlike small animals; gastropods, which include snails and their relatives; bivalves, which are clams and other animals with two shells; and cephalopods, which consist of octopuses and squid. Chitons, gastropods, and bivalves are common along the coast.

All mollusks share some common traits. The anatomical structures of a typical bivalve are shown in Figure 4.1. The bodies of most mollusks are covered with one or two shells. The functions of shells include protection from predators and desiccation as well as points of attachment for muscles. Each mollusk has a soft body with internal organs that carry out circulation, respiration, reproduction, digestion, and excretion. The body is covered with a thin tissue called a mantle. In some species the mantle secretes the shell as well as one or more defensive chemicals, such as ink, mucus, or acid. A mollusk also possesses a muscular foot that is used for locomotion, whether it be swimming, digging, or crawling. Except for animals in the bivalve group, each mollusk feeds by scraping up food with a file-like tongue, the radula. This muscular organ can efficiently take in algae, animal tissue, or detritus. The bivalves are filter feeders that trap and consume food particles that are suspended in water.

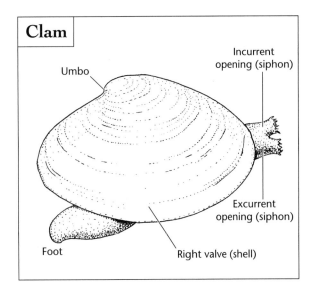

Fig. 4.1 The anatomy of a clam is typical of all bivalves: Its shell is secreted by a thin layer of tissue, the mantle. A muscular foot provides locomotion.

In mollusks, sexes are separate. Bivalves release their eggs and sperm into the water, and fertilization occurs there. In other types of mollusks, internal fertilization, a process in which sperm are transferred to the body of the female, is the norm. Following internal fertilization, females deposit strings or cases of eggs on the sand, seaweed, or rocks. In both internal and external fertilization, zygotes develop into swimming larvae that metamorphose into adult forms.

The eight-segmented, flattened body of a chiton (see the upper color insert on page C-3) attaches to rocks with a strong, muscular foot. If dislodged from its rock, a chiton will curl into a tight ball, very much like the defensive position of its terrestrial look-alike, the pill bug. Chitons are, for the most part, herbivores that graze on marine algae during high tide, scraping it off of rocks with their sharp radula. The red chiton (*Ischnochiton ruber*), which measures about 0.5 inch (1.3 cm) long, has reddish valves that are dotted and dashed with white and brown markings. The bee chiton (*Chaetopleura apiculata*), with its white or cream-colored valves and scattered, short hairs, grows slightly larger, reaching lengths of 0.75 inch (1.9 cm).

The gastropods, perhaps some of the best-known mollusks, include snails, limpets, and nudibranchs. Most have one shell, although some species have none. A gastropod's head is equipped with sensory organs, including eyes that are light-sensitive dots, as well as tentacles and a mouth. Locomotion is provided by one large, flat foot located in the center of its body. Many species of snails and limpets have shells that contain and protect the animals' internal organs. In some, an operculum protects the occupant from danger.

Limpets are close relatives of snails. They have uncoiled, conical shells that are similar in appearance to the typical Chinese straw hat. Nearly all limpets live on coastal rocks, where they erode the rock to create indentions in which to hide. When the tide is in, limpets leave their dens and graze on algae growing on rocks, scraping it off with their file-like tongues. Many have ornate shells with multiple swirls, although the shell of one of the most common species—the 1-inch (2.5-cm) long tortoiseshell, or Atlantic plate limpet (*Acmaea testudinalis*)—is a simple cone.

Despite their common name, the keyhole limpets are not true limpets, although they are closely related mollusks. Keyhole limpets can be distinguished from true limpets by small holes in the tops of their shells. To feed, keyhole limpets pull water in under their shells, send it over their gills, and then expel it through the holes in their shells. *Diodora cayenensis*, a keyhole limpet of the Atlantic coast that grows to 7 inches (17.7 cm), feeds primarily at night.

Snails are slow-moving mollusks whose bodies are protected by spiral shells, many of which can be closed with a thin, pale operculum. Members of one type, the turban shell snails, each have a thick, dark-colored operculum that resembles a flattened pearl. The black turban snail (*Tegale funebralis*) feeds on diatoms and algae growing on rocks.

Moon snails are fierce carnivores that dine on mollusks with two shells, such as clams and mussels. When hunting, a moon snail plows through the wet sand, digs out its prey, and holds it tightly with its strong foot. The snail secretes a fluid that coats and softens a small area of the prey's shell, then the snail drills a perfectly round hole through the softened spot with its radula. To feed, the snail sucks its prey out of the hole. Moon snails lay eggs on the beach in fragile packets generally known as sand collars. Several species are common in intertidal zones, including the Atlantic moon snail (*Policices duplicatus*), milk moon snail (*Polinices lacteus*), and northern moon snail (*Lunatia heros*). The Atlantic oyster drill (*Urosalpinx cinera*) feeds on oysters using a dissolving and drilling technique very much like moon snails.

Snails of the genus *Murex* are familiar gastropods on shores. At one time, murex snails were harvested because the fluid produced by a gland in their feet was used as a dye. Murex snails have ornate shells with several spikes and spires, which they use to pry open the shells of clams and other two-shelled prey. The radula is then inserted into the open shell, where it rips out pieces of flesh.

The whelks are yet another group of snails that feed on mussels, clams, and their relatives. Whelks lay eggs in long strips of capsules, each holding up to 100 eggs. Near the top of each capsule there is a small opening that serves as an escape route for the larvae when they hatch. The New England neptune (*Neptuna decemcostata*), a 4-inch (10-cm) long whelk, has a white shell that is banded with brown spirals. The lightening whelk (*Busycon contrarium*) is about 6 inches (15 cm) long and spends much of its time buried in sand. The Atlantic dog whelk (*Thais lapillus*), about 1.5 inches (3.8 cm) long, was once prized by Native Americans for its purple pigment.

Periwinkles, snails with cone-shaped, whorled shells, are widespread in intertidal zones. The common periwinkle (*Littorina littorea*) browses on films of algae along the shore. When the tide is out, mucus produced by the snail's foot seals the animal to a rock or other solid substrate. The snail withdraws into its shell and closes its operculum to retain moisture, then waits until the tide returns. Unlike most marine snails, the common periwinkle can breathe for a short time out of water. Its mantle is rich in blood vessels and serves as a site for gas exchange with the air.

Slipper shells, such as the eastern white slipper shell (*Crepidula plana*), are gastropods whose uncoiled, lipped shells resemble one-half of a mussel or clam shell. Even though this species possesses the typical gastropod foot and is capable of movement, it does not graze algae like most other snails, so it rarely moves. Instead, a slipper snail collects food by suspension feeding, a technique in which water enters the shell on one side, passes over the gills, and exits the shell on the other side. Plankton in the water sticks to the mucus-coated gills, and the radula scrapes it off and transfers it to the mouth. To reproduce, slipper shells form stacks made of several animals, with the oldest ones on bottom and the youngest ones on top. Young animals are always male but can change sex if the population needs more females for reproductive purposes. During this transformation, the male reproductive organs wither and female organs develop.

A conch is a large, herbivorous snail that uses its tubelike proboscis to suck up meals of algae. If startled, a conch can launch its body forward and out of harm's way, using its operculum as a springboard. The heavy shell of a conch is rimmed with thick, flaring lips. One of the largest species, the queen conch (*Strombus gigas*), grows to 1 foot (30 cm) in length. The hawk-wing conch (*Strombus raninus*) and Florida fighting conch (*Strombus alatus*) are close relatives, but smaller, measuring about 3 inches (7.5 cm) long.

Several types of gastropods, such as the opalescent sea slug in the lower color insert on page C-3, do not possess shells. The nudibranchs, or sea slugs, are shell-less gastropods that also lack proboscises, although they do possess radulae and jaws. *Nudibranch* literally means "naked gills" and refers to the fact that the sea slug's gills are not covered with a shell. Some have appendages on their backs called cerata, clearly visible on the Taylor's sea slug in the upper color insert on page C-4, which act as supplemental gills. One species, the striped nudibranch (*Cratena pilata*) is gray or green, with wide white margins and rusty brown stripes. The sea lemon (*Onchidoris muricata*), a yellow or cream-colored nudibranch, has a ring or whorl of gill-like filaments around the anus.

Sea hares, another group of gastropods without shells, are characterized by their cauliflower-shaped gills that are protected by flaps of skin. Wide extensions of a sea hare's foot, as well as tall antennae, give the animal an appearance similar to a hare. Most species, such as the spotted sea hare (*Aplysia dactylomela*) and Willcox's sea hare (*Aplysia willcoxi*), are herbivores that can be found crawling or swimming among seaweed.

Unlike gastropods, which have one shell or lack shells completely, bivalves are mollusks that possess two shells. Bivalve shells, or valves, hinge together on one side and are opened and closed by strong muscles. Clams, mussels, scallops, and oysters are some of the many types of bivalves found in the intertidal zone.

The body of a bivalve, which is safely tucked between two hard, protective shells, does not have a head. A large foot extends through partially opened shells to attach to a substrate or to burrow into the sand. In species that burrow, a section of the mantle is modified to form two siphons, tubes that extend from the body to bring water to the gills. In a filter-feeding bivalve, gills do more than function in respiration; they also trap food. The gills of filter feeders, which are located in the mantle cavity, are covered with mucus that ensnares bits of food in the water and cilia that guide the food toward the animal's mouth.

The soft-shell clam (*Mya arenaria*) burrows deep into the sand, extending its long siphons up in the water. Unlike some other species, the soft-shell clam's siphons will not completely retract into its shells when threatened by predators. The Florida coquina (*Donax variabilis*), whose siphons are shorter, can pull its siphons inside its shells and close them tightly. Because its siphons are short, this clam cannot burrow very deep into the sand. The Atlantic surf clam (*Spisula solidissima*) has heavy, thick-lipped shells that are visible at low tide. The hard-shell clam (*Mercenaria mercenaria*), whose shell is colored purple on the inside, buries itself just below the surface of sand. Jackknife clams (*Ensis directus*), often found in mudflats, are speedy bivalves that can burrow through the sand faster than most predators can dig them out.

Unlike most bivalves, scallops lie on the surface of the sand instead of burrowed in it. They have no need for, and do not possess, a muscular foot. The adult bay scallop (*Aequipecten irradians*) can swim by rapidly closing its shells and jetting water out near the hinge. Juveniles spend their time clinging to shallow water plants to stay out of silty sediment that could clog their gills. In both juvenile and adult stages, the edges of the mantles are fringed with rows of bright blue eyes whose primary function is to detect movement.

Another bivalve that lives on top of the sand, the mussel is commonly found attached to rocks or other solid substrates. To secure its body in place, a gland near the mussel's foot produces tough strands of byssus, a slightly elastic protein. The shell of the horse mussel (*Modiolus modiolus*) is easily identified by its distinctive bushy, black coat. The blue mussel (*Mytilus edulis*) is found growing in beds or colonies on all types of substrates, including rocks, piers, pilings, and even mud. By forming mussel carpets, these animals are able to protect themselves from wave energy by dispersing it over a large area. When the tide goes out, these smooth-shelled animals respire by circulating moist air over their gills.

Oysters may be one of the best-known bivalves along the coast. An oyster cements one valve to a rock or other solid substrate and spends its entire life in that place. Oyster shells are heavy, rough, and variable in shape. Unlike most other bivalves, the oyster lacks a foot and siphons, so must open its shell to circulate water over its gills. The common oyster (*Crassostrea virginica*), like those seen in the lower color insert on page C-4, is found along much of the eastern coast of North America.

A cockle is a heart-shaped bivalve that burrows into the sand with its muscular foot. The foot can also be used to spring from one location to another, a good strategy for escaping predators. The great heart cockle (*Dinocardium robustum*), with its yellow shell accented with brown marks and 35 ridges, or ribs, grows to 5 inches (12.5 cm) in width. Morton's egg cockle (*Laevicardium mortoni*) is a small, smooth-shelled species that is yellow with purple or brown marks. Nicknamed the duck clam, this cockle is one of several types sought by diving ducks.

Despite their names, shipworms (belonging to the genus *Teredo* or the genus *Bankia*) are not worms at all, but bivalves that bore into wood. A young shipworm uses its small, file-like shell to dig a tunnel in wood, then climbs inside. Throughout its life, the bivalve continuously enlarges the structure to accommodate its growing body, which can reach lengths of 1 foot (30 cm). On one end, the unusually shaped bivalve has a foot and a two-valved shell. On the other end are its siphons and two shell-like extensions, called pallets, that are used to seal the open end of the tunnel. Shipworms remain in their tunnels their entire lives, feeding on the wood and nutrients filtered from water.

Arthropods

Arthropods make up an extremely large group of animals in the intertidal zone. Their terrestrial cousins are the familiar, ever-present insects and spiders. Along the coast, typical marine arthropods include crabs, shrimp, lobsters, and horseshoe crabs.

The body of an arthropod is covered with a hard shell called the exoskeleton, which provides structural support and protection from predators. The skeletons of arthropods are primarily composed of chitin, an extremely tough but highly flexible material made of long chains of molecules that are similar in structure to cellulose.

Arthropods' bodies are divided into segments. The head, which represents the first segment, is a region that contains the brain, many sensory organs, and specializations for taking in food. Among the arthropod's sensory organs are compound eyes, which create multiple pictures and arrange them like tiles in a mosaic, and antennae, which are organs of touch.

Arthropods, whose name translates to "joint legged," are able to move quickly because they have appendages that are jointed. An appendage is a leg, antenna, or other part that extends from a segment of the body. Appendages are used for a variety of functions, including food handling, walking, swimming, and sensory input.

Advantages and Disadvantages of an Exoskeleton

More than 80 percent of the animal species are equipped with a hard, outer covering called an exoskeleton. The functions of exoskeletons are similar to those of other types of skeletal systems. Like the internal skeletons (endoskeletons) of amphibians, reptiles, birds, and mammals, exoskeletons support the tissues and give shape to the bodies of invertebrates. Exoskeletons offer some other advantages. Serving as a suit of armor, they are excellent protection against predators. Also, because they completely cover an animal's tissues, exoskeletons prevent them from drying out. In addition, exoskeletons serve as points of attachment for muscles, providing animals with more leverage and mechanical advantage than an endoskeleton can offer. That is why a tiny shrimp can cut a fish in half with its claw or lift an object 50 times heavier than its own body.

Despite all their good points, exoskeletons have some drawbacks. They are heavy, so the only animals that have been successful with them over time are those that have remained small. In addition, an animal must molt, or shed, its exoskeleton to grow. During and immediately after a molt, an animal is unprotected and vulnerable to predators.

In most arthropods, sexes are separate. Mating occurs at prescribed times of the year, and courtship rituals are often complex. In many species the male deposits sperm in the female's body. The sperm are held there until eggs mature, then as each egg leaves the ovary, sperm are released and fertilization occurs. Resulting zygotes mature into larvae that swim in the plankton for a short period of time before settling down on the seafloor to mature.

Crustaceans

Crustaceans make up the largest group of arthropods in the intertidal zone. Crabs, shrimp, and lobsters, as well as many small animals that often go unnoticed, are classified as crustaceans. The body segments of crustaceans are grouped into three specialized areas: the head, the thorax, and the abdomen. In many organisms, the head and thorax are fused and protected by a section of the exoskeleton called the carapace. The head is equipped with five pairs of appendages: Two sets are antennae, and three sets are used for feeding. Some species of crustaceans have claws that are large and capable of exerting hundreds of pounds of pressure. The thorax and abdomen of a crustacean possess walking or swimming appendages. Sexually, most types of crustaceans have separate genders, and females are in charge of egg laying and brooding.

Barnacles are sessile crustaceans that live within protective plates. Positioned so that their heads are down and their legs are sticking out into the water, barnacles spend their lives attached to wood, rock, or some other solid surface. At low tide, when they are exposed to air, barnacles close their plates to prevent drying. These plates have a small opening that allows moist air to circulate over their gills, providing oxygen.

Unlike most of their close relatives, barnacles are hermaphrodites, but individuals usually cross-fertilize. One barnacle can transfer sperm to its neighbor by way of a long, extensible tube. Fertilized eggs develop into zygotes that later form swimming larvae. Eventually, the larvae settle on a substrate, exude a brown glue that anchors them in place, and metamorphose into the adult form.

Acorn barnacles, a group of acorn-shaped crustaceans, are widespread in intertidal waters. The adults are shrimplike animals contained within overlapping, protective plates, as shown in Figure 4.2. When covered with water, the plates open and six pairs of feathery handlike appendages, called cirri, extend. The cirri search for bits of food floating in the water. Colors and sizes of acorn barnacles vary from species to species. The northern rock barnacle (*Balanus balanoides*) is white but may look green if overgrown with algae. The little gray barnacle (*Chthamalus fragilis*) is small, about 0.5 inch (1.2 cm) and dingy gray in color.

Goose barnacles are closely related to acorn barnacles but barely resemble them. The two-part shell of a goose barnacle is supported by a fleshy stalk. The striped goose barnacle (*Conchoderma virgatum*) reaches a height of 3 inches (7 cm). This species sometimes washes ashore or is found growing on stranded whales. The pelagic goose barnacle (*Lepas*), identified by its purple, rubbery stalks and bright white plates, grows to $\frac{3}{4}$ inch (2 cm) and can be found attached to almost any floating substrate, including wood and tar.

Amphipods (belonging to the genera *Gammarus* or *Talorchestia*) are tiny crustaceans known by a variety of common names such as beach or sand hoppers, scuds, and beach fleas. Found on beaches and in tide pools, these buglike animals have five pairs of legs for walking and three pairs for swimming, as well as appendages that help them jump long distances. Many of these amphipods are filter feeders that live burrowed in tubes in the sand, while a few species scavenge detritus.

The isopods are another group of small crustaceans. Like their terrestrial cousins, the pill and sow bugs, isopods will roll into a ball if disturbed. Most crawl around on the substrate, although they are capable of swimming. The habitat, size, and diets of different groups of isopods vary greatly. The sea pill bug (*Sphaeroma*) lives among seaweeds in tidal pools. The tiny gribble (*Limnoria*), only 0.5 inch (0.6 cm) in length, feeds on fungi that grow on wood. This miniature isopod bores into driftwood as well as docks to secure its meals.

Shrimp, lobsters, and crabs are crustaceans that are closely related to one another. They have five pairs of legs, the first of

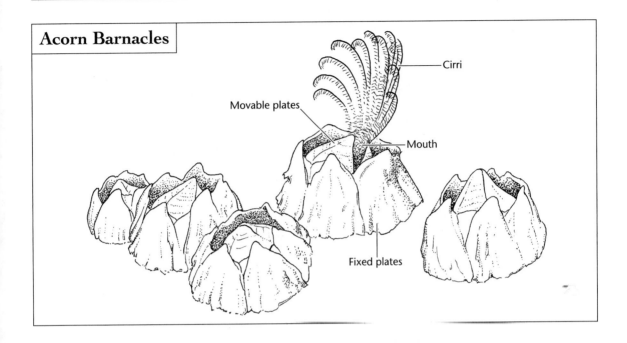

Fig. 4.2 *An acorn barnacle lives inside a permanent home of protective calcium plates.*

which is often modified as claws. The term *shrimp* is not a scientific one but refers to a group of small, 10-legged crustaceans that have relatively light-weight exoskeletons and gills located under their shells. Most get around by swimming rather than crawling.

Snapping, or pistol, shrimp of the family Alpheidae resemble lobsters, although they are much smaller, most growing to lengths of 1.5 to 2 inches (3.8 to 5 cm). The front margins of their exoskeletons extend forward, partially covering their eyes. All snapping shrimp possess one gigantic claw that, in some species, is longer than the shrimp's body. When it is snapped closed, the claw makes a loud sound that is used to warn away intruders. One of the most common shallow water types, shore shrimp (*Palaemonetes*) spend most of their time buried in sand or sediment rather than swimming.

The sand shrimp (*Crangon septemspinosa*) is the small cousin of the familiar, edible prawn, an inhabitant of deeper waters. Growing to only 2.5 inches (6.3 cm), this little shrimp is easy to identify because the front of its exoskeleton is shortened. The long-eyed shrimp (*Ogyrides alphaerostris*) is an intertidal resident whose eyes sit atop long, slender stalks. The animal's body is transparent, but the appendages show spots of red and green color.

In the lobster family, the northern, or American, lobster (*Homarus americanus*) pictured in the upper color insert on page C-5, is the only intertidal species. Adults can grow to 3 feet (90 cm) in length and weigh up to 45 pounds (20 kg), although most weigh in at close to 1 pound (0.45 kg). The northern lobster lives among rocks on the shoreline where it hunts slow-moving animals. When this is not an option, the lobster can scavenge, using its antennae to locate dead animals floating in the intertidal zone. If alarmed, this crustacean executes a strong tail flip to propel its body backward through the water.

To reproduce, a male lobster transfers its sperm to a female who has recently molted. The sperm is stored in the female's body until she produces eggs. As the eggs pass out of her body, they flow over the reservoir of sperm and are fertilized. Fertilized eggs stick to the female's small swimming appendages, where they incubate for the next 10 to 12 months. The spiny, swimming hatchlings molt several times before becoming bottom-dwelling adult forms.

Several groups of animals are described as crabs. Like all crustaceans, crabs have hard, protective shells. Their bodies are flattened, an adaptation that allows them to squeeze into small spaces between rocks where they are protected from waves and predators. Crabs can stay out of water for a long time, getting oxygen from the air. Most crabs are active, aggressive animals that prey on smaller organisms and scavenge anything they can find.

The true crabs have four pairs of walking legs, some of which have been modified as paddles for swimming (see Figure 4.3). The blue crab (*Callinectes sapidus*), olive or blue green on its dorsal side with bright blue claws, averages about 9 inches (22.5 cm) in width. Much of the time, the blue crab can be found buried in sand, with only its eyes exposed to the surface.

Another true crab, the green crab (*Carcinus maenas*) is green on the dorsal side, but its ventral color varies by sex. Males of this species are yellow ventrally while females are red orange. Green crabs are typically found hiding under rocks and in seawalls in the intertidal zones.

Ghost crabs (*Ocypode quadrata*), shown in Figure 4.4, are small, sand-colored members of the true crab group that are familiar sights to most beachcombers in the southern United States. Their gills are able to extract oxygen from the air, enabling these animals to stay out of water much of the time. Ghost crabs dig 2- to 4-feet (60- to 120-cm) deep burrows that slope downward at a 45-degree angle in the sand. Occasionally, they emerge to rush to the water where they wet their gills and look for bits of food that may have washed ashore.

Hermit crabs are distinguished from true crabs by their two pairs of walking legs. Hermits are well known for their portable living arrangements. Lacking a hard exoskeleton over its abdomen, a hermit crab finds an abandoned snail shell and moves in, backing its soft, twisted abdomen into the curved interior space. Inside the shell, two pairs of small legs hold the animal in place. If threatened, a hermit crab fully retreats into its shell and closes the opening with a claw.

Fig. 4.3 The bodies of crabs are divided into segments: the head and the thorax, which are fused by the carapace, and the abdomen. Crabs have four pairs of legs and a pair of claws that serve as pincers.

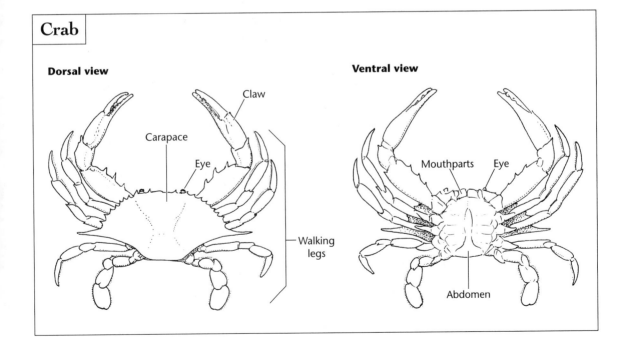

Fig. 4.4 Ghost crabs live in burrows on sandy beaches. (Courtesy of America's Coastline Collection, NOAA)

Several species of hermit crabs are common along the coast. Reaching a maximum length of 2 inches (5 cm), the long-clawed hermit crab (*Pagurus longicarpus*) prefers the shells of periwinkles and oyster drills. The slightly smaller flat-clawed hermit crab (*Pagurus pollicaris*), with its broad, pink claw, prefers to live in the shells of whelks and moon snails.

Sea Spiders and Horseshoe Crabs

Sea spiders and horseshoe crabs are two close relatives of crustaceans that have misleading common names. Unlike true spiders whose bodies are made of two segments, sea spiders have three distinct body parts: a head, a trunk, and a short abdomen. The head bears a proboscis, a mouth, two feeding appendages, and four eyes. Walking legs are located on the trunk. These tiny animals, measuring only 0.08 inch (2 mm) in length, blend in with seaweeds, which help camouflage them. The female sea spider, the larger of the sexes, lays eggs, which are externally fertilized by the male. After fertilization, the male glues the eggs to his legs to protect them until they hatch. Most species feed on sessile animals such as sponges and soft corals by tearing away pieces of flesh, piercing them with their proboscis, and sucking out the juices. Common coastal species include the ringed sea spider (*Tanystylum orbiculare*), lentil sea spider (*Anopladactylus lentus*), and anemone sea spider (*Pycnogonum littorale*).

The horseshoe crab, pictured in the lower color insert on page C-5, is actually more closely related to land spiders than to crabs. The horseshoe-shaped shell ends in a long tail that serves as a rudder as the animal swims. Two compound eyes and two simple eyes are located on the dorsal side of the shell. Underneath the shell are five pairs of legs, the first four modified for walking and the last pair for swimming. The gills resemble folded pages and are located near the posterior end. Horseshoe crabs feed on burrowing mollusks and worms.

During mating, the smaller male hangs onto the female's back, attaching himself with special hooks located on his first pair of legs. At the shoreline, the female digs a hole and deposits her eggs, which are immediately fertilized by the male. The hatchlings are $\frac{3}{4}$ inch (1.2-cm) larvae that look like miniature adults.

The Atlantic horseshoe crab (*Limulus polyphemus*) is a regular visitor to many coasts between Maine and the Gulf of Mexico. The adults travel to shallow water in the spring and lay eggs. As they mature, the young gradually move to deeper water. The females are 2 feet (60 cm) long, including the tail.

Echinoderms

Starfish (sea stars), sea cucumbers, brittle stars, sand dollars, and sea urchins are members of a group of spiny-skinned animals known as echinoderms. All echinoderms are radially symmetrical, and most have five or more arms extending from a central disk, as shown in Figure 4.5. Their bodies are supported by internal skeletons which are solid in some species and jointed in others. Most echinoderms are designed so that their mouths are located on the ventral sides and their anuses on the dorsal surface. Depending on the species, the nutritional strategies of echinoderms vary from carnivores to detritivores or herbivores.

Tentacle-like structures called tube feet help echinoderms travel slowly across the intertidal zone as well as grasp prey. The tube feet act as suction pads, clutching and releasing surfaces as they move across them. This is possible through a water vascular system that supplies water through small muscular tubes to individual suction cups. As the tube feet press against an object, water is withdrawn, creating suction. When water is returned to the cups, the suction is broken and the tube feet release their grip.

To reproduce, echinoderms release sperm and eggs into water, where they fuse to form zygotes. Larvae swim in the plankton for a short time, then settle to the bottom and take on typical echinoderm features. Most echinoderms can also reproduce asexually. If part of the animal breaks off, a piece may grow into a complete, new organism. All are capable of regenerating missing limbs, spines, and in some cases, intestines.

Fig. 4.5 The tube feet of a starfish are visible on its underside. (Courtesy of Dr. James P. McVey, NOAA Sea Grant Program)

Starfish and brittle stars are two groups of echinoderms that generally stay in one location for long periods of time but occasionally crawl slowly over rocks and sand looking for prey. During extremely turbulent weather, they withstand strong incoming waves by wrapping their bodies around rocks. If caught by a predator, a brittle star's arm breaks off easily, allowing it to escape. The lost arm regenerates in a short time. The spiny brittle star (*Ophiothrix angulata*) and short-armed brittle star (*Ophioderma brevispina*) are typical intertidal species.

Starfish, such as the ones in the upper color insert on page C-6, are predators that enjoy feeding on mussels. As the tide goes out, a starfish crawls into a mussel bed, selects a victim, then pulls the prey's shells apart with its strong legs. As soon as the shell opens, even a fraction of an inch, the starfish turns its body inside out and pushes its stomach through its own mouth and into the opening between the shells. Digestive juices in the starfish's stomach dissolve the mussel tissue, which the stomach then absorbs. When the meal is over the echinoderm puts its stomach back in place and crawls to another mussel. *Asterias vulgaris*, the northern sea star, is a mussel-eating species that is common in intertidal waters. *Henricia leviuscula*, the blood star, uses an entirely different technique for feeding. This species is a suspension feeder that traps food particles on the mucus it exudes atop its body. Cilia quickly sweep the trapped food into the blood star's mouth.

Tennis ball–sized spined spheres found resting in intertidal waters are echinoderms called sea urchins (see the lower color insert on page C-6). Ten sections of rigid shell that extend from the mouth on the ventral side to the anus on the dorsal side, like sections of an orange, support their bodies internally. The mouth contains a feeding structure, Aristotle's lantern, made of five teeth that come together in a beaklike arrangement. The urchin uses these teeth to scrape algae from rocks.

Seaweed is the favorite food of the purple sea urchin (*Arbacia punctulata*), whose body is covered in 1-inch (2.5-cm) purple spines. During low tide, sea urchins use their tough spines to carve out holes in the rocks. Once inside their caves, they are protected from desiccation during low tide as well as from crashing waves when the tide rolls in. During low tides, the holes retain water, providing the moisture that sea urchins must have to survive. Sometimes, the sea urchins grow too large to get out of their caves, and become prisoners in them. They survive by extending their tube feet through the openings to catch bits of algae.

The much larger long-spined sea urchin (*Diadema antillarum*) is a Florida resident that hides during the day and searches for algae at night. Small shrimp and fish may live among its spines, taking advantage of its protection. The green sea urchin (*Strongylocentrotus droebachiensis*) has shorter spines than the long-spined sea urchin, although its body may be slightly larger. Both coastal species are commonly found in tidal pools and on rocky ledges.

Sand dollars are so flat that at first look they do not resemble their echinoderm relatives. The plates of their skeletons are fused and fixed, and the external surfaces are covered with tiny spines that look like fuzz or a short coat of hair. Sand dollars use their spines to burrow into sand. Once buried, the spines collect small food particles that fall among the spines and transfer them to the mouth. As shown in the upper color insert on page C-7, the common sand dollar (*Echinarachnius parma*) may be brown with purple or red tints. A five-petal pattern on the sand dollar's dorsal side corresponds to the five legs of a starfish. Holes near the tips of the petals allow the animal to extend tube feet, which are used for respiration. Most specimens measure about 3 inches (7.5 cm) in width.

Sea cucumbers are a group of tubular echinoderms that lie on their sides with their mouth at one end of the tube and anus at the other end. Around the mouth, several tube feet are modified to form tentacles. Some species use these to catch bits of food suspended in water or to pick up food-laden sediment. The respiratory structures of these animals are located alongside the digestive tract.

The number and arrangement of tube feet on a sea cucumber vary, depending on the species. Usually found clinging to intertidal rocks, the orange-footed sea cucumber (*Cucumaria frondosa*) has tube feet arranged in five distinct bands along the length of its body. However, only the ones on the ventral side possess suckers. On the other hand, the body of the hairy sea cucumber (*Sclerodactyla briareus*) is covered with tube feet.

Conclusion

Mollusks, crustaceans, and echinoderms are large groups of invertebrates that demonstrate many evolutionary advances over the systems of such simple invertebrates as sponges, cnidarians, and worms. Complex respiratory, digestive, circulatory, excretory, and reproductive systems cater to the needs of all cells within these large organisms. In many, the soft body parts are protected with shells or skeletal tissue.

Mollusks commonly found in the intertidal zone include bivalves and gastropods. Bivalves are soft-bodied animals that are protected by two strong shells. Their bodies are covered by a delicate tissue, the mantle, which secretes the shell and defensive chemicals. Most are filter feeders that trap small bits of food in the mucus on their gills. Clams, mussels, and oysters are examples of coastal bivalves.

Intertidal gastropods include a number of snails such as moon snails, black turbans, and oyster drills. Gastropods are generally predators that feed on bivalves. A gastropod feeds by climbing atop a bivalve, secreting juices on the shell, then boring through the shell with its sharp radula. Once a hole is created, the gastropod floods the soft body of its prey with digestive enzymes and then sucks out the soupy mixture.

Crustaceans are members of the arthropod group, the most successful animals on Earth. The bodies of these invertebrates are segmented and protected with tough exoskeletons. On the coast, crustaceans include buglike amphipods and isopods, shrimp, lobsters, and crabs. All of these animals are equipped with appendages for walking, swimming, or both. In most, fertilization is internal, and fertilized eggs are brooded on legs or swimming appendages. Close relatives to the crustaceans are the sea spider and horseshoe crab, animals that share some characteristics with terrestrial spiders.

Echinoderms are known as spiny-skinned animals because their protective outer layer is tough and prickly. The bodies of these animals are radial, and most are divided into five equal parts. Sea stars, brittle stars, sea urchins, sand dollars, and sea cucumbers are examples of echinoderms. All members are capable of slow movement on their hydraulically controlled tube feet.

As a group, intertidal invertebrates are highly adapted to the problems of life in shallow water and on shifting substrates. They exhibit a variety of mechanisms that protect them from desiccation and predators, as well as modifications that keep them from being washed out to sea by crashing waves and changing tides. No other group of invertebrates on the Earth lives in a more challenging environment.

Coastal Fish
Life in Shallow Seawater

Once or twice each day, seawater floods across the intertidal rocks, mudflats, and sands. As it recedes, some is trapped in rocky depressions and low spots, creating tiny pockets of water called tide pools. A tide pool is a miniature seawater ecosystem that supports an astonishing variety of living things, including small bony fish. The largest group of vertebrates in the sea, fish are well represented in the intertidal zone.

The different species of fish that occupy tide pools share some commonalities. All are highly adapted to life in an area of the sea where water rushes over them during the incoming tide then recedes, often leaving them uncovered and vulnerable to dehydration. The fish of the tide pools are generally small, with large heads and elongated bodies. Their bodies lack some of the structures of open-water fish, like swim bladders. Foods for tide pool fish include crustaceans, insect larvae, algae, worms, and amphipods. The fish that occupy the tide pools represent many different genera and species, but as a general rule they belong to one of five major taxonomic families: Cottidae (sculpins), Pholidae (gunnels), Gobiidae (gobies), Blennidae (blennies), or Gobiesocidae (clingfish).

Sculpins

The Cottidae family includes about 300 different species of fish, with most members being marine, although a few are found in freshwater. The majority of species, such as the one in Figure 5.1, are bottom dwellers that live in either temperate or Arctic waters. Most members of this family have large heads with eyes located high up on the head, elongated bodies, and pectoral fins that look like fans. Some are partially covered with scales, while others are completely scale-less. Their pelvic fins have two to five soft rays as well as a single spine; their anal fins are spineless. Members of this family usually lay eggs close to rocky areas, and the males are in charge of guarding the eggs until they hatch.

The tide pool sculpin (*Oligocottus maculosus*) is one member of the Cottidae family that makes its home in the intertidal zone. Geographically, sculpins have a very wide range, including the North and South Pacific and the Atlantic Oceans. This little fish rarely grows larger than about 3 inches (7.5 cm) in length and lives a maximum of five years. Its green or red dorsal side is decorated with five irregular dark circles while the ventral surface is typically blue green accented by a white line. The head is somewhat large with a blunt snout, and the body is elongated, tapering to a point at the end of the tail.

Like all sculpins, tide pool sculpins are highly adapted for their environments and have developed a variety of structures

Fig. 5.1 Sculpins are small fish that can be found in rocky tide pools. (Courtesy of Historic NMFS Collection, NOAA)

▲ At low tides, patterns formed by tidal currents can be seen in the sand. *(Courtesy of Olympic Coast National Marine Sanctuary, NOAA)*

◀ Anemones live among rocks in shallow intertidal waters. *(Courtesy of NOAA Damage Assessment and Restoration Program)*

▲ *The red body fluid of a bloodworm can be seen through its pale skin.* (Courtesy of National Estuarine Reserve Collection, NOAA)

▲ *Sabellid worms are polychaetes that live in parchment-like tubes. They exchange gases and gather food with their feather tentacles.* (Courtesy of Dr. James P. McVey, NOAA Sea Grant Program)

▲ *A chiton clings to a rock with its muscular foot.* (Courtesy of Mary Hollinger, NOAA Biologist, NODC)

▲ *The opalescent sea slug, a nudibranch, is a shell-less mollusk.* (Courtesy of National Estuarine Research Reserve Collection, NOAA)

▲ *A Taylor's sea slug crawls over the sandy bottom of shallow intertidal water.* (Courtesy of National Estuarine Research Reserve Collection, NOAA)

▲ *Oysters form extensive beds in both tidal and subtidal waters.* (Courtesy of Bob Williams, U.S. Fish and Wildlife)

▲ *The northern lobster, or American lobster* (Homarus americanus), *is a crustacean with two strong pincers.* (Courtesy of Rich Wahle, Damage Assessment Restoration Program, NOAA)

▲ *During spawning, the smaller male horseshoe crab* (Limulus polyphemus) *hooks himself to the larger female.* (Courtesy of Mary Hollinger, NOAA Biologist, NODC)

▲ *A variety of starfish, or sea stars, make their homes in the intertidal waters of rocky coasts.* (NOAA National Estuarine Research Reserve Collection)

▲ *Rock-boring sea urchins use their spines to carve out holes in rocks where they can be protected from the waves crashing on shore.* (Courtesy of NOAA, Sanctuary Collection)

▲ *Sand dollars use their short spines to burrow into the sand.*
(Courtesy of NOAA National Estuarine Research Reserve Collection)

▲ *A Galápagos sea lion and a marine iguana share a warm spot on the shore.* *(Courtesy of Rosalind Cohen, NODC, NOAA)*

▲ *Sandpipers wade in the surf looking for prey.* (Courtesy of Mary Hollinger, NESDIS/NODC biologist, NOAA)

▲ *Walrus use their large canines to crush the shells of the invertebrates on which they feed.* (Courtesy of Captain Budd Christman, NOAA Corps)

that help them survive challenges such as displacement by energetic water, predation, fluctuating temperatures, and oxygen-poor conditions. Strong incoming tides can carry the fish from one tide pool to another, but they can navigate back to the tide pools of their birth by using chemical tracking. Structures that sense chemicals, called chemoreceptors, enable the sculpins to identify and differentiate the chemicals that make each tide pool unique.

The ability to change colors to match their surroundings helps tide pool sculpins blend with the environment, enabling the fish to avoid many predators and providing them with camouflage while they wait to ambush their own prey. Their diet consists of crabs, shrimp, and the eggs of insects and other fish.

For the tide pool sculpins, spawning occurs from December through February. The female deposits her eggs on rocks or on sand in the intertidal regions. During one breeding season, a female may lay eggs several times. Males deposit sperm, which externally fertilize the eggs. Eggs incubate for a couple of weeks before hatching into young fish that spend about a year in the juvenile stage. The immature sculpins hide in very shallow pools or small pockets of water that are tucked behind rocks and shielded from the onrushing waves.

The wooly sculpin (*Clinocottus analis*) is a species that inhabits tide pools all along the Pacific coast. These fish, which occur in a host of colors including gray, green, brown, or red, grow from 3 to 7 inches (7.5 to 17.8 cm) in length. Wooly sculpins have contrasting white, yellow, or pink markings on the surface of their bodies and are covered in prickles. Their heads and upper jaws possess whiskerlike tentacles called cirri.

Never straying very far from one location, a wooly sculpin lives in a small world, within a radius no greater than 3 or 4 feet (0.9 to 1.22 m). This small fish spends a lot of time resting quietly on the bottom of the tide pool, where it blends in with the surroundings. The wooly sculpin is a good hunter that can lunge quickly through the water when worms, snails, or small crabs come within range. Its broad pectoral fins can propel the animal upward and forward with speed and accuracy. Showing

Bony Fish Anatomy

All bony fish share many physical characteristics, which are labeled in Figure 5.2. One of their distinguishing features is scaly skin. Scales on fish overlap one another, much like shingles on a roof, protecting the skin from damage and slowing the movement of water into or out of the fish's body.

Bony fish are outfitted with fins that facilitate maneuvering and positioning in the water. The fins, which are made of thin membranes supported by stiff pieces of cartilage, can be folded down or held upright. Fins are named for their location: Dorsal fins are on the back, a caudal fin is at the tail, and an anal fin is on the ventral side. Two sets of lateral fins are located on the sides of the fish, the pectoral fins are toward the head, and the pelvic fins are near the tail. The caudal fin moves the fish forward in the water, and the others help change direction and maintain balance.

Although fish dine on a wide assortment of food, most species are predators whose mouths contain small teeth for grasping prey. Nutrients from digested food are distributed through the body by a system of closed blood vessels. The circulation of blood is powered by a muscular two-chambered heart. Blood entering the heart is depleted of oxygen and filled with carbon dioxide, a waste product of metabolism. Blood collects in the upper chamber, the atrium, before it is pushed into the ventricle. From the ventricle, it travels to the gills where it picks up oxygen and gets rid of its carbon dioxide. Water exits through a single gill slit on the side of the head. The gill slits of fish are covered with a protective flap, the operculum.

In many bony fish, some gases in the blood are channeled into another organ, the swim bladder. This organ is essentially a gas bag that helps the fish control its depth by adjusting its buoyancy. A fish can float higher in the water by increasing the volume of gas in the swim bladder. To sink, the fish reduces the amount of gas in the bladder.

Most bony fish reproduce externally. Females lay hundreds of eggs in the water, then males swim by and release milt, a fluid containing sperm, on the eggs. Fertilization occurs in the open water, and the parents swim away, leaving the eggs unprotected. Not all of the eggs are fertilized, and many that are fertilized will become victims of predators, so only a small percentage of eggs hatch.

Fig. 5.2 The special features of bony fish include bony scales (a), opercula (b), highly maneuverable fins (c), a tail with its upper and lower lobes usually of equal size (d), a swim bladder that adjusts the fish's buoyancy (e), nostrils (f), pectoral fins (g), a pelvic fin (h), an anal fin (i), lateral lines (j), dorsal fins (k), and a stomach (l).

Coastal Fish 87

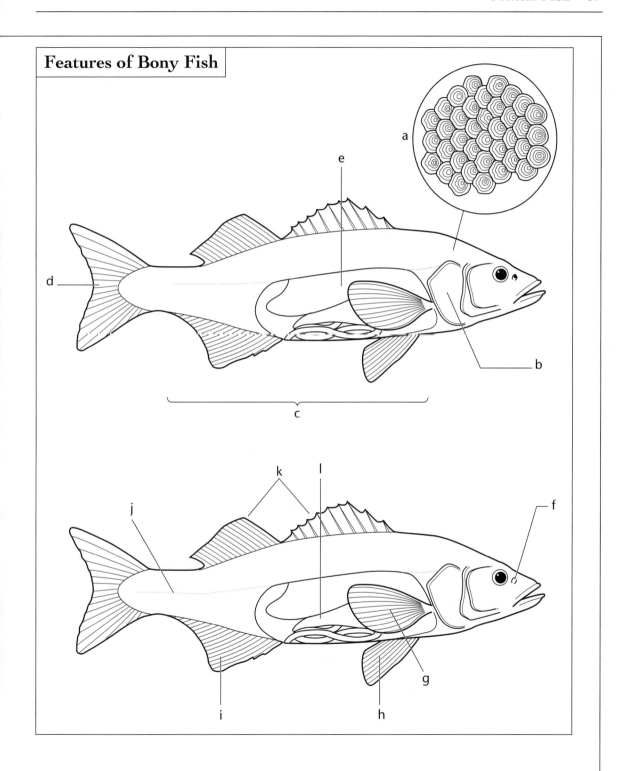

a remarkable adaptation for a fish, the wooly sculpin can use those same pectoral fins to climb out of the tide pool when oxygen levels are dangerously low. Out of the water, the fish can breathe air for up to 24 hours.

Wooly sculpins spawn all year long, but the peak season is from November to May. The female deposits her eggs in nests where they are fertilized by males. Unlike many other tide pool fish, neither sculpin parent stays behind to guard the eggs, which generally hatch within 13 to 18 days.

Gobies

Found among the sculpins and other intertidal residents are members of the family Gobiidae. Worldwide this is the largest

Colorization

One of the most striking features of fish is their colorization. Coloring and body marks on fish help them avoid predators by staying out of sight. Many prey species, such as gulf flounder, avoid being eaten by blending in with their surroundings, matching the subtle shades of their habitats. Spotted fish look like the seafloor, and striped fish blend in with grasses. Some reef fish display bright colors because they live among brightly colored sponges and corals.

Conversely, coloring that mimics fish's habitats helps predators get close to their prey. The ability to avoid detection is a significant advantage for such hunters as scorpion fish that wait quietly until prey comes within striking distance. A hunter is able to conserve both time and energy if it does not have to pursue its food.

Most fish, including sea trout and grouper, display some degree of countershading. This form of coloring reduces the clarity of the fish's body outline in water. The simplest, and most common, form of countershading is a dark dorsal side and a pale ventral side, with intermediate colors between the two. When sunlight filters down through the water, it lightens the fish's back and throws shadows on its underside. The overall effect of countershading lessens the degree of contrast between the fish and the water.

A few species of fish, such as the spotfin butterfly fish and the high hat fish, show disruptive or deflective colorization that

family of marine fish, boasting about 2,100 different species that are found almost everywhere except in polar water. Populations of this family are most concentrated in the tropics and subtropics, but some species even survive in subarctic streams of Siberia.

Members of the Gobiidae family, commonly called the gobies, are small fish with long bodies. Adults range from 1 inch (2.5 cm) to 20 inches (50.8 cm) in length. The scales of the gobies are small, and their backs are equipped with two dorsal fins. The first dorsal fin has eight flexible spines, and the second fin is soft. Some species are vividly colored, while others are marked in camouflaging shades of drab grays and browns. Like most intertidal fish, gobies have blunted, round heads and large eyes. Their pelvic fins are fused together to form a

includes bands, stripes, or dots of contrasting colors. These colors and patterns confuse predators by distorting the true shape, size, and position of the fish. Bright patterns draw the predator's eye, causing it to see the pattern rather than the fish itself. This type of coloring can deflect the predator's attention away from a fish's vulnerable areas, such as its head and eyes.

Colorization can also be used as an advertisement. There is no point in being poisonous and unpalatable if no one knows it. Instead of hiding, poisonous fish announce their dangerous status. Fish may also advertise their age or sex with coloring. Males are generally more colorful than females, whose duller shades help camouflage and protect them. Young fish may be transparent or pale, making it hard for predators to spot them, as well as letting the older fish of their own species know that they are not a threat.

Fig. 5.3 Scorpion fish lie quietly on the seafloor, camouflaged by their colors and patterns. (Courtesy of the Fisheries Collection, NOAA)

structure that looks and functions very much like a suction cup. Gobies depend on these structures to help them cling to rocks and other surfaces in times of violent wave action.

The male goby is in charge of readying the nest for the female to lay her eggs. Once eggs are deposited and fertilized, he stays to watch and defend them. Gobies feed on a wide variety of items including crustaceans, mollusks, worms, sponges, small fish, and the eggs of other fish or invertebrates. They are secretive fish, rarely straying far from their burrows.

There are many species of gobies that make their home among the tide pools, but one of the most familiar inhabitants is the common goby (*Pomatoschistus microps*). This tiny, sandy-colored fish, which measures only 2.5 inches (6.35 cm) long at maturity, blends in perfectly with the bottom of the tide pool. The common goby is found all along the British coast, as far south as Portugal and as far north as southern Norway. In mating season, the female may lay multiple batches of eggs. The eggs are deposited for safety purposes under a shell, and the male goby stands guard the 11- to 14-day period until the small fish hatch. During their short (15 months) but active lives, common gobies prey on small crustaceans and worms.

Blennies

The third major family of fish occupying the tide pools, Blennidae, includes about 300 different species of fish. Members of this family, commonly referred to as blennies, share many of the traits of gobies but have a few notable differences. Whereas gobies have small scales over their body, blennies have scale-less skin that is covered with mucus. Also, because of their ability to skillfully jump from one tide pool to another, never falling short of their target or touching the dry rocks in between, blennies are nicknamed "rock skippers."

The common blenny (*Lipophrys pholis*) occupies the beach and intertidal zone. Average size for the common blenny is about 6 inches (15.2 cm), and coloration is usually green with black molting, although some are gray and black with red-rimmed eyes. Members of this species prefer rocky areas below the low-water mark and mid-shore zones where they hide under the edges of rocks or in crevices between rocks in pools. Like the sculpin, the common blenny can crawl out of oxygen-deficient pools by hoisting its body up on rocks or weeds. When frightened, the fish hops immediately back into its tide pool.

At breeding time the male changes color, becoming completely black except for a white mouth and a pale blue fringe that adorns the tip of the long dorsal fin. Breeding occurs during the spring in very shallow water. After eggs are laid and fertilized, the male of the species stands guard. Male blennies are fierce protectors, running off would-be predators and even sharply biting the fingers of curious humans who stick their hands in tide pools.

The teeth of a common blenny are sharp and comblike, ideally designed for scraping barnacles off rocks and for crunching crabs into bite-size pieces. Other small invertebrates are also important sources of food. Blennies have to be on constant watch against their own predators, which include larger fish as well as seabirds.

The zebra blenny (*Istiblennius zebra*), which is about 6 inches (15.3 cm) long at maturity, makes its home in the intertidal zone of the Hawaiian Islands. The body color varies from blue to yellow, but it is the distinctive dark, vertical stripes along the fish's side that account for its common name. The zebra blenny uses its pelvic fins as props to support its head high in the water, a position that enables the animal to watch for predators and prey. Distinctive, tentacle-like cirri extend down over its eyes. This tiny fish can curl its tail to one side like a coiled spring, then forcefully unfurl it to jump into a nearby tide pool.

A favorite food of the zebra blenny is seaweed, which it scrapes from rocks with comblike teeth. As with other blennies, the male prepares a nest then uses bright colors and specific mating behaviors to attract a female to his nest site. The male watches the eggs until they hatch about a week later.

The Skin and Senses of Fish

Most marine fish have two protective layers covering their bodies: skin and scales. All fish are covered with skin, a defense against infection by bacteria and fungi. The skin is also a site of gas exchange, enabling fish to augment the supplies of oxygen obtained through their gills. Other body systems that incorporate the skin are the excretory, which consists of organs that eliminate wastes in the body, and the sensory, which collects information about the environment and sends that information to the brain for processing.

Depending on the type of fish, skin may contain glands that produce mucus or poison. Mucus on the skin's streamlined surface reduces the amount of friction, or drag, a fish experiences as it swims. Poison produced and secreted by skin is an excellent deterrent against predators.

Some of the sensory structures on skin include receptors to detect taste, touch, and vibrations. Whiskerlike skin outgrowths called barbels act as fingers, enabling a fish to feel around for prey that might be hidden below sand or sediment. In some types of fish, the barbels contain receptors of taste and smell.

Most fish have scales over their skin. Bony fish are generally covered with very thin, flexible scales that can move independently, allowing the fish to have plenty of maneuverability. Scales are only attached at one end, where they are embedded in the dermis. The opposite end is completely free so that each plate can slide and shift across others when the fish bends and flexes.

All fish are born without scales. Scales begin to form as the young fish matures, although some species remain scale-less their entire lives. Scales of most fish are made of two layers: a bony layer and a fibrous layer. Their shape, size, and arrangement are often used to help classify fish, as well as determine their age. As a fish matures, its scales grow, forming growth rings in the process. Rings on scales can be counted very much like the rings on a tree.

A fish depends of its senses of olfaction (smell) and taste to detect chemicals that are dissolved in the water. Smell helps a fish pick up clues about prey, even when it is distant. Taste is most often used to sample the source of the smell. Taste buds can be located on the skin, on barbels, and on the tongue. The degree of sensitivity to taste depends on the type of fish and its environment; for example, a fish in murky water might be more dependent on chemical

Gunnels

The fourth major family of fish of the intertidal zone is Pholidae, commonly known as the gunnels. The gunnels are found hiding under edges of rocks and in crevices in tide

sensory input than one that is swimming in clear water.

For many marine fish, smell is the primary sense. The olfactory structures are located in the head of a fish, and they sample water from the outside environment by pulling it in through openings called nasal apertures. Water flows into the nasal capsule, a structure that contains olfactory sensory cells.

Fish that find their way back "home" are able to detect the unique chemicals of the waters where they were born. Their brains may have been stamped with a sensory imprint when they were young, and the fish may retain that imprint as they age. For the rest of their lives, homing fish recognize the chemical signatures of their birthplace.

Although olfaction is the primary sense used by most fish, some species also depend on their sense of sight. Fish that live in clear water are much more light dependent than fish that live in turbid or deep, dark waters. Keen eyesight can help capture prey, locate a mate, and receive warnings. Since colors of light are scattered and refracted as they travel through water, only those fish in very shallow water are exposed to many colors.

Hearing in fish is processed in the inner ears, organs that receive input from vibrations created by sound. Fish do not have external ear canals like those in humans that connect the inner ear to the external environment. Sounds travel well in water and in fish's bodies, where they cause small bones in the inner ears to vibrate. The bones are denser than the rest of the animals' tissues and therefore respond differently to sound waves. As the wavelengths of sounds vary, so do their effects on the small bones. These differences are registered by nerves and are interpreted as sound.

Not all types of fish experience the same sensitivities to sound. One variable to sensitivity is the position of the swim bladder. The swim bladder is filled with gas, so its density is not the same as the density of the rest of a fish's body tissue. When sound waves strike the swim bladder, it vibrates, stimulating tissues that are connected to it. These vibrations aid the fish in interpreting sounds. Species that lack swim bladders are not as sensitive to sound as those who possess them. The fish that are most sensitive to sound are those whose swim bladders are directly connected to the middle ear.

pools in cold waters along the northern Pacific and northern Atlantic Oceans. Colors among the 15 different species include bright orange, lime green, and drab shades of brown or gray. The largest gunnel grows to 18 inches (45.7 cm), but most Pholidae species are much smaller, generally around

3 inches (7.6 cm) in length. The bodies of gunnels are eel-like with very long dorsal fins and flexible spines. The pelvic fins are either very tiny or completely absent. Gunnels prefer to dine on worms, crustaceans, amphipods, algae, and insect larvae. Sometimes during the winter season they do not eat at all.

The penpoint gunnel (*Apodichthys flavidus*) is one of the longest species. This fish, a giant in the intertidal zone, abounds in the rocky coastal zones between Alaska and southern California. Running from the rear of its head all the way to the intersection of the caudal fin is a long dorsal fin. Color varies from lime green to red brown, with a distinctive black bar radiating from the eye. Sharp, pointed teeth are perfect for dining on small crustaceans and mollusks.

The penpoint gunnel spawns from December through March depending on its geographic location. After the white, round eggs are laid in a cluster, one or both of the parents coil around the mass to incubate them. Incubation lasts about 10

Water Balance

All living things must deal with the problem of maintaining homeostasis, or a stable internal chemical environment. Living cells are very sensitive to the relative amounts of water and salt in their environment. Changes in internal and external levels of salinity can disrupt normal cellular functions.

Compared to an equal volume of fish tissue, seawater has a higher concentration of salt molecules and a lower concentration of water molecules. Water has a natural tendency to flow from an area where it is more highly concentrated to one where it is less concentrated. In marine fish, water tends to flow out of the tissues, leaving the body through the gills, digestive system, and excretory system. For this reason, marine fish are always losing water, putting them at risk of dehydration. To maintain as much water as possible in their tissues, marine fish produce very little urine.

To replace the water that is constantly leaving their bodies, marine fish continuously drink seawater. As they do so, they also consume a lot of dissolved salts. These are absorbed into the intestine, transported to the gills, and then eliminated. Salts do not naturally flow out of fish's bodies, so they must be actively transported by special excretory cells.

weeks, and the transparent hatchlings are about 0.5 inch (1.2 cm) long when they emerge.

Another intertidal zone gunnel is the butterfish (*Pholis gunnellus*). A layer of slippery slime that makes it impossible to hold this squirming little creature in one's hands for long accounts for its common name. The butterfish has an eel-like body, causing it to closely resemble a small snake; in fact, when it moves, it wriggles across the bottom of the tide pool in a very snakelike motion. Identification of this species is not difficult as butterfish have obvious orange-brown coloration and 13 large, distinct spots on the back. This colorful little fish, which may reach lengths of up to 10 inches (25.4 cm) but is usually much shorter, feeds on crustaceans and worms that find their way to the tide pool.

Clingfish

The final major family of fish living in the intertidal zone is Gobiesocidae, or the clingfish. Clingfish get their common name from the way they hang on to anything available to avoid being washed out to sea. Clingfish can form suction cups from their pelvic fins and use those devices to secure their bodies to seaweed or rocks when the high tide washes over them. In fact, the clingfish is so good at holding to rocks that it is sometimes accidentally carried out to sea when strong waves dislodge rocks from the shoreline.

The suction cup of the clingfish serves more than one purpose. Not only does it enable the fish to hang on, but it also holds water when conditions are dry. After the tide goes out, the water level in a tide pool can fall dangerously low. If this happens, the fish uses the moisture held within the suction cup to breathe. Sheltered beneath a rock or within the crevice of a rock, it takes in oxygen from the trapped moisture and waits until the tide rolls back in with a fresh supply of water, oxygen, and nutrients.

Members of the family Gobiesocidae are considered to be ray-finned fish because their fins have one spine and four or five rays. Each fish in the family has a single dorsal fin that is positioned nearer to the tail than to the head, as well as an anal

Territoriality

Fish living close together may show territorial behavior, the tendency to occupy and defend an area, usually to eat and reproduce there. There are many different patterns of territorial behavior. Some species are territorial all the time, but others may only display this behavior during reproductive periods. Depending on the situation, fish may be territorial against their own species, toward other species, or both.

Territorial behavior requires a lot of energy, and a fish cannot afford to expend more energy on defending its territory than it takes in as food. For this reason, fish have developed several threatening displays that involve a lot of posturing and ritualized motions yet conserve energy. If an intruder gets close to a damselfish's alga garden, for example, the damselfish first attempts to scare it away with a threatening posture of spread fins and gill covers. If this strategy does not solve the problem, the damselfish makes excited, aggressive movements. Only as a last resort will it attempt to chase away an intruder, as actual chase could lead to a fight that might end in the death of the defender.

fin. The head is broad and flat, and the body is smooth, scale-less, and extremely slick owing to a thick coat of mucus.

This large family of fish, which includes 35 different genera and 120 species, is found in both temperate and tropical waters of all oceans. The largest member of this family measures about 12 inches (30.5 cm), but the majority of Gobiesocidae members are about 3 inches (7.62 cm) in length.

In many ways, the northern clingfish (*Gobiesox maeandricus*) is representative of the group. Also known as the flathead, or common clingfish, this species lives along the rocky shores from Alaska to Baja California. Most of the day the fish rests in tide pools or hides under rocks. The largest northern clingfish ever measured was 6.5 inches (16.5 cm) long, but most are much smaller. Their primary foods are crustaceans, worms, and mollusks, which they eat with their sharp, pointed teeth.

Spawning season is dependent on the exact location of this species and can range from November through May. The female deposits her eggs under rocks or shells in the intertidal zone. After she leaves, the male guards them over the next several weeks until the little ones hatch.

Conclusion

Fish in the intertidal zone are highly adapted for life in the shallow, energetic water. Most seashore fish are members of five fish families, the sculpins, gunnels, gobies, blennies, and clingfish. Typically, beach and intertidal zone fish have large heads, long thin bodies, and adaptations for holding on to rocks or other substrates. In most species, after females lay eggs in the shallow water, males are in charge of watching and guarding them.

Sculpins are small but conspicuous fish in tide pool environments. A sculpin only grows to about 3 inches (7.5 cm) in length, and its coloration is either green or red with five irregular dark circles. The head is large in proportion to the rest of the slim, elongated body. If washed out of its home tide pool, a sculpin can use chemoreceptors to identify the specific chemical combinations that mark its home and then eventually make its way back to it.

Gobies are very small fish that are found worldwide. Their pelvic fins are modified to form suction cups that anchor them during times of strong wave action. Clingfish, like gobies, can also arrange their fins in suction cups that help them avoid being displaced by strong waves. If the oxygen levels drop significantly, clingfish can climb out of the water and breathe air for a few hours. To do so, these fish use water stored in the pectoral fins to keep the gills moist.

Blennies, known as rock skippers because they are able to jump from one tide pool to another, also live in tide pools. During the breeding season, male blennies change color and carry out ritualized breeding activities. After the eggs are laid, the male protects them until they hatch.

The eel-like bodies of gunnels have long dorsal fins and flexible lateral fins, so they swim back and forth in very much the same movement as snakes. The butterfish, a gunnel that can grow to be 10 inches (25.4 cm) long, is coated with mucus, an adaptation that most likely developed to reduce drag in water.

All species of tide pool fish face unusual environmental conditions and possess special adaptations to help them survive in this unique situation. Like other vertebrates in the ocean, fish are consumers that feed near the top of the food chain and important members of shoreline ecosystems.

6

Reptiles, Birds, and Mammals
Vertebrates at the Edge of the Ocean

The seashore is home to a wide range of vertebrates, animals with backbones. Many of the vertebrates grow much larger than the invertebrates. In the seashore food chains, vertebrates fill the roles of consumers. The group includes amphibians, reptiles, birds, and mammals, most of which are air breathers that spend only part of their lives in the water. With the exception of the amphibians, all classes are represented on or near the edge of the ocean.

Only vertebrates that spend the majority of their lives on the coast are classified as coastal residents. The beach and intertidal zones are home to very few reptiles; for example, although all sea turtles visit shorelines to lay their eggs, they do not live there. The only full-time coastal reptile is the marine iguana. Many birds are true residents, spending their lives wading in shoreline waters in search of food, while thousands of others occasionally visit or feed along the intertidal

---

Body Temperature

Animals that are described as warm blooded, or endothermic, maintain a constant internal temperature, even when exposed to extreme temperatures in their environment. In mammals, this internal temperature is about 97°F (36°C), while in birds, it is warmer, around 108°F (42°C).

Warm-blooded animals have developed several physiological and behavioral modifications that help regulate body temperature. Since their bodies generate heat by converting food into energy, they must take in enough food to fuel a constant body temperature. Once heat is produced, endotherms conserve it with insulating adaptations such as hair, feathers, or layers of fat. In extreme cold, they also shiver, a mechanism that generates additional heat.

Heart rate and rate of respiration in warm-blooded animals does not depend on the temperature of the surroundings. For this reason, they can be as active on a

zone. Likewise, a variety of mammals visit shorelines, but a relatively small number of them are full-time coastal species.

Marine Reptiles

Although several types of turtles visit the shoreline, the area serves as the primary home to only one reptile, the marine iguana (*Amblyrhynchus cristatus*). In fact, the marine iguana is given the distinction of being the only seagoing lizard in the world. This ominous-looking, but docile creature lives in intertidal zones along the rocky shores of the Galápagos Islands, which are located in the equatorial waters west of Ecuador. Populations of marine iguanas are large, and scientists estimate that about 300,000 of these reptiles live there. This figure translates to 4,500 of these organisms per square mile of the islands, making them the most numerous animal on the Galápagos Islands.

Marine iguanas, like the one in the lower color insert on page C-7, range in length from 10 inches (25.4 cm) to up to 2 feet (0.61 m). Most are dark gray or black, although some males have green or red blotches on them. The noses of marine iguanas are short and blunt, and their front teeth, or cuspids, are razor sharp, a feature that makes it easy for them to scrape algae

---

cold winter night as they are during a summer day. This is a real advantage that enables warm-blooded animals to actively look for food year round.

The internal temperature of cold-blooded, or ectothermic, animals is the same as the temperature of their surroundings. In other words, when it is hot outside, they are hot, and when it is cold outside, they are cold. In very hot environments the blood temperature of some cold-blooded animals can rise far above the blood temperature of warm-blooded organisms. Furthermore, their respiration rate is dependent on the temperature of their surroundings. To warm up and speed their metabolism, cold-blooded animals often bask in the sun. Therefore, cold-blooded animals such as fish, amphibians, and reptiles, tend to be much more active in warm environments than in cold conditions.

Marine Reptile Anatomy

Reptiles are not usually associated with marine environments. In fact, of the 6,000 known species of reptiles, only about 1 percent inhabits the sea. Members of this select group include lizards, crocodiles, turtles, and snakes. Each of these organisms shares many of the same anatomical structures that are found in all reptiles: They are cold-blooded, air-breathing, scaled animals that reproduce by internal fertilization. Yet, to live in salt water, this subgroup has evolved some special adaptations not seen in terrestrial reptiles.

In turtles, the shell is the most unique feature. The lightweight, streamline shape of the shell forms a protective enclosure for the vital organs. The ribs and backbone of the turtle are securely attached to the inside of the shell. The upper part of the shell, the carapace, is covered with horny plates that connect to the shell's bottom, the plastron. Extending out from the protective shell are the marine turtle's legs, which have been modified into paddle-like flippers capable of propelling it at speeds of up to 35 miles per hour (56 kph) through the water. These same legs are cumbersome on land, making the animals slow and their movements awkward.

Most air-breathing vertebrates cannot drink salty water because it causes dehydration and kidney damage. Seawater contains sodium chloride and other salts in concentrations three times greater than blood and body fluids. Many marine reptiles drink seawater, so their bodies rely on special salt-secreting glands to handle the excess salt. To reduce the load of salt in body fluids, these glands produce and excrete fluid that is twice as salty as seawater. The glands work very quickly, processing and getting rid of salt about 10 times faster than kidneys. Salt glands are located on the head, often near the eyes.

There are more than 50 species of sea snakes that thrive in marine environments. Sea snakes possess adaptations such as nasal valves and close-fitting scales around the mouth that keep water out during diving. Flattened tails that look like small paddles from rocks. Long, very sharp claws enable the lizards to cling to the rocks and avoid being swept away when they are pounded by the surf.

Each marine iguana has a white covering atop its head that looks similar to a wig. This white cap is not due to pigmentation in the skin but rather is a crusty layer of salt. Because they consume a lot of salty seawater as they feed on algae, the bodies of marine iguanas have a strategy to lower the amount of salt in their systems. A gland located between the nostrils

easily propel these reptiles through the water. The lungs in sea snakes are elongated, muscular air sacs that are able to store oxygen. In addition, sea snakes can take in oxygen through the skin. Their adaptations to the marine environment enable sea snakes to stay submerged from 30 minutes up to two hours; however, this ability comes at a cost. Because marine snakes routinely swim to the surface to breathe, they use more energy and have higher metabolic rates than land snakes. To balance their high energy consumption, they require more food than their terrestrial counterparts.

Finally, crocodiles usually occupy freshwater, but there are some species that live in brackish water (in between salt water and freshwater) and salt water. These animals have salivary glands that have been modified to excrete salt. Their tails are flattened for side-to-side swimming and their toes possess well-developed webs. Saltwater crocodiles are equipped with valves at the back of the throat that enable them to open their mouths and feed underwater without flooding their lungs.

and eyes stores salt that is taken in during feeding. When the gland is filled near capacity with salt, the lizard holds its head upward and releases a mighty sneeze. The liquefied salt shoots high into the air and falls back on the head of the lizard, where it later hardens into a white crust.

The diet of marine iguanas primarily consists of nine different species of algae, all of which grow underwater and in the tidal zone of the islands. At times they will also ingest grasshoppers, crustaceans, and the afterbirth of sea lions.

Normally, feeding occurs only once a day. Mature lizards swim out into the pounding surf to dive for algae, while young members of the group eat algae growing on rocks in the tidal zone.

Marine iguanas, like other reptiles, are ectotherms, so they function best when the environmental temperature is about 95°F (35°C). However, the water surrounding the Galápagos Islands is extremely cold, and a dive can cause an iguana's internal body temperature to plunge to 50°F (10°C). For this reason, marine iguanas must take special measures to warm themselves. After feeding, each lizard takes a sunbath by flattening their bodies on dark rocks, exposing as much skin surface as possible to the sun. The sunlight brings their body temperature back up to the normal range, and it also helps their digestive systems work efficiently, since they perform better at high temperatures. At night, temperatures on the Galápagos Islands drop dramatically, so lizards sleep in large huddles to conserve heat.

When the weather is warm, marine iguanas can get too hot, so they also take precautions to cool themselves. One cooling technique is to position their bodies so that only a minimum of surface is exposed to full sunlight. The lizards also lift their bodies to allow cool air to circulate beneath them, creating a cooling breeze. If this technique is not enough to reduce their body temperatures, they can always retreat to shady areas between cracks in rocks.

Marine iguanas are not solitary creatures and usually live peacefully in large colonies. Only during mating season do males display aggressive behavior to one another. The season for mating generally lasts from December through March, and females only mate one time per season. A female is sexually mature between three and five years of age, while a male matures between six and eight years of age.

Just prior to mating season, males develop vibrant colors that distinguish them from the drab-colored females. It is not uncommon for male iguanas to engage in a ritual of head bobbing to establish their territory and warn other males away. If the warning does not work, two male iguanas engage in a contest where they butt heads, attempting to push the other backward. The contest may last up to five hours and ends when one male walks away or becomes submissive to the other male.

After the male and female lizards mate, the female stays with her colony for about a month. When it is time to lay eggs, she leaves the group and searches for a soft, sandy area. She selects a spot anywhere from a few yards up to several miles away from the colony where she spends about a half day digging her nest. The female lays from one to six eggs and stands guard over them for the next 16 days before making her journey back to the colony. Incubation requires between 89 and 120 days, and when the infant lizards hatch, they scurry across the sand in search of protective cover.

The primary predators of marine lizards are hawks, herons, frigate birds, and short-eared owls. In recent years, as the human populations of the Galápagos Islands have grown, humans, dogs, and pigs have also become predators of the lizards.

Marine lizards usually have ample food, but certain conditions can reduce the availability of algae. In the past, El Niño phenomenon, which causes ocean water around the Galápagos Islands to warm, has killed the algae species that the marine iguana eats. The only algae that survived the event were brown algae, which are toxic to marine iguanas, so many lizards died. El Niño is not a yearly event, however, and in most years lizards feed safely.

Seabirds

The coast supports hundreds of different species of birds, many of which spend the majority of their lives on or near the shoreline. Some members of this gigantic group are waders that walk about the edge of the ocean looking for food, while others fly from one location to the other, rarely getting their feet wet.

It is not unusual to see flocks of shorebirds using their long bills to explore the soil along the water's edge. Many of the different types of birds on beaches and in intertidal zones feed on the same kinds of prey—worms and insect larvae—but

Marine Bird Anatomy

Birds are warm-blooded vertebrates that have feathers to insulate and protect their bodies. In most species of birds, feathers are also important adaptations for flying. As a general rule, birds devote a lot of time and energy to keeping their feathers waterproof in a process called preening. During preening, birds rub their feet, feathers, and beaks with oil produced by the preen gland near their tail.

The strong, lightweight bones of birds are especially adapted for flying. Many of the bones are fused, resulting in the rigid type of skeleton needed for flight. Although birds are not very good at tasting or smelling, their senses of hearing and sight are exceptional. They maintain a constant, relatively high body temperature and a rapid rate of metabolism. To efficiently pump blood around their bodies, they have a four-chambered heart.

Like marine reptiles, marine birds have glands that remove excess salt from their bodies. Although the structure and purpose of the salt gland is the same in all marine birds, its location varies by species. In most marine birds, salt accumulates in a gland near the nostrils and then oozes out of the bird's body through the nasal openings.

The term *seabird* is not scientific but is used to describe a wide range of birds whose lifestyles are associated with the ocean. Some seabirds never get further out into the ocean than the surf water. Many seabirds are equipped with adaptations of their bills, legs, and feet. Short, tweezerlike bills can probe for animals that are near the surface of the sand or mud, while long, slender bills reach animals that burrow deeply. Several styles of seabird bills are shown in Figure 6.1. For wading on wet soil, many seabirds have lobed feet, while those who walk through mud or shallow water have long legs and feet with wide toes.

Other marine birds are proficient swimmers and divers who have special adaptations for spending time in water. These include wide bodies that have good underwater stability, thick layers of body fat for buoyancy, and dense plumage for warmth. In swimmers, the legs are usually located near the posterior end of the body to allow for easy maneuvers, and the feet have webs or lobes between the toes.

All marine birds must come to the shore to breed and lay their eggs. Breeding grounds vary from rocky ledges to sandy beaches. More than 90 percent of marine birds are colonial and require the social stimulation of other birds to complete the breeding process. Incubation of the eggs varies from one species to the next, but as a general rule the length of incubation correlates to the size of the egg: Large eggs take longer to hatch than small ones do.

Bills of Shorebirds

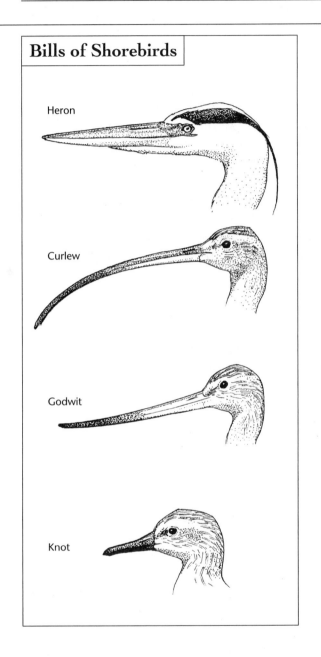

Fig. 6.1 *The bills of shorebirds are adapted for a variety of feeding styles.*

avoid competing with one another by collecting their food from different parts of the substrate. Long-billed shorebirds, such as oystercatchers, can probe deep into the soil to extract burrowing worms and bivalves such as oysters. In one year, a single oystercatcher can consume about 100 pounds (45.4 kg) of oysters. The shorter bills of sandpipers (see the upper color insert on page C-8) are very sensitive and flexible and are perfect for finding prey in the upper levels of soil. The bills of snipe and similar birds are designed to probe the sand and locate prey by sense of feel, then suck the meal directly into the birds' mouths.

The noisy birds referred to as oystercatchers belong to the family Haematopodidae. Their common name comes from their ability to chip oysters and other mollusks off the rocks along the shore. Their strong bills are also capable of prying open clams, oysters, and mussels to get to the soft meat inside the shell. When these delicacies are not available, oystercatchers dine on worms and other invertebrates that they pluck from underneath the sand.

Oystercatchers can be found along coasts worldwide except in the polar regions. Many have black-and-white bodies with pale legs, although there are some species that are completely black or dark brown. Of the 10 species of oystercatchers, six are commonly seen along the beach and rocky coasts around the world.

Sandpipers are members of the Scolopacidae family, shorebirds that are found worldwide except in Antarctica. During nonbreeding times of the year, they congregate in large flocks on beaches. Most sandpipers breed in the Northern Hemisphere and use yearly migration routes that follow shorelines. They make their nests by digging shallow depressions in the soil or sand, then lining them with grasses. The eggs, typically four, are incubated by the female or by both parents, depending on the species.

There are as many as 80 different species of sandpipers and their characteristics vary, especially relating to body and bill lengths; however, coloration in all of these wading birds is fairly uniform, with most having grayish brown backs and light-colored bellies. Their sizes range from 5 inches (12.7 cm) to 24 inches (60.96 cm), including the bills, which can be short, long, and straight or long and curved. All shapes and sizes are used to probe shallow water for crustaceans, mollusks, or worms.

A commonly seen beach and intertidal member of this family is the least sandpiper (*Calidris minutilla*), which measures only about 5 inches (12.7 cm) in length. In some locales this species is known as the peep because of the distinctive sounds it makes while running along the beaches searching for tiny food items left exposed by the receding tide. The least sandpiper spends its winters along the Atlantic coast and breeds in Alaska and Canada.

The third major family of shorebirds, Charadriidae, is better known as the plovers. These small birds, with their long wings and pigeonlike bills, are found globally except for Antarctica. The Charadriidae family contains 65 different species, making plovers a very diverse group. Many species spend the majority of their time hopping along the beach or seashore, but there are some species that make their home far inland from the beaches.

Plovers nest on the ground on open, sparsely vegetated sand or gravel beaches. The female lays three or four eggs in a shallow depression that she has lined with pebbles and shell fragments. The eggs are well camouflaged, but there are times when a hungry predator may stumble across them. To defend its nest, an adult plover performs a convincing broken-wing act. As the predator approaches, the parent abandons the nest and pretends to sit on a nest in a new location, hoping to lure the predator away from the real one. If this routine does not work, the bird fakes an injury by dragging one wing along the ground as if it is broken. As the predator follows, the bird continues the act, always moving away from the nest and staying just out of reach of its pursuer. When the nest appears to be safe, the plover flies away and returns to its young. One species that does an excellent broken-wing act is the killdeer (*Charadrius vociferus*), named for the "kill-dee" cry for which it is known.

All but one species of beach-dwelling plover snatch insects and tiny crustaceans from the ocean waves as the tide recedes from the shore. The ruddy-colored turnstone (*Arenaria interpres*) feeds by turning over shells, pebbles, and seaweed to find crustaceans and insects hiding below. Plovers breed in the high Arctic tundra around the North Pole and migrate southward to North America, Africa, Asia, New Zealand, and Australia.

The most familiar shorebirds are members of the family Laridae, a group known as the gulls. Except for a few tropical areas, all the world's beaches provide homes for some species of gulls. There are at least 50 different species, and most of these are very large, solidly built, and web footed.

A typical adult gull is gray and white, but head markings and colors of feet and bills vary with species. The younger gulls tend to be black or brown in color. Gulls are not picky eaters and will consume just about anything that comes their way. They are known as scavengers that sift through human garbage piles as well as the trash they find along beachfronts. Gulls also steal eggs out of the nests of other birds. Sometimes gulls rely on their own hunting ability to catch fish or shellfish for dinner. Their bills are not designed to open crabs and other shellfish, so they carry them aloft, drop them on rocks or pavement, then return to the ground to find the soft delicacies within.

Fig. 6.2 The herring gull is a common sight on North American seashores. (Courtesy of Mary Hollinger, NESDIS/NODC biologist, NOAA)

The gulls are equally at home on land, at sea, and in the air. They are adept at walking on land but are talented fliers and good swimmers. Gulls build their nests near the sea but have been found living inland, on offshore islands, in busy cities, and at garbage dumps.

The most common gull of North America is the herring gull, or seagull (*Larus argentatus*), pictured in Figure 6.2. The adult is silver gray on its back and white on its belly, and reaches 22 inches (55.8 cm) in length. The tips of the herring gull's wings are primarily black with some assorted white spots. The legs and feet are pink, and the yellow bill is decorated with a bright red patch. Herring gulls, unlike many of the gull species, do not venture more than a few miles from land. They nest in large colonies throughout Alaska and Canada, on the Atlantic coast, and even inland toward the Great Lakes of North America. The birds spend winters in Mexico or the West Indies.

Mammals

The aquatic mammals of the coasts are carnivores that belong to the suborder Pinnipedia. Members of this suborder, commonly referred to as the pinnipeds, or "fin-footed" animals, use flippers for locomotion on land and in the water. Their sizes range tremendously, from relatively small animals that weigh a mere 200 pounds (90.7 kg) to organisms that are as heavy as several tons. Pinnipedia can be divided into three main families: Phociadae (true seals), Otariidae (eared seals, which includes sea lions and fur seals), and Odobenidae (walruses).

There are noticeable differences between the three families, but pinnipeds share some commonalities. All members have adaptations that allow them to move with agility through the water, although they are slow and clumsy on land. Members of Pinnipedia have streamlined bodies that lack external ears

Marine Mammal Anatomy

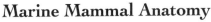

Mammals are warm-blooded vertebrates that have hair and breathe air. All females of this group have milk-producing mammary glands with which to feed their young. Mammals also have a diaphragm that pulls air into the lungs and a four-chambered heart for efficient circulation of blood. The teeth of mammals are specialized by size and shape for particular uses.

Marine mammals are subdivided into four categories: cetaceans, animals that spend their entire lives in the ocean; sirenians, herbivorous ocean mammals; pinnipeds, web-footed mammals; and marine otters. Animals in all four categories have the same characteristics as terrestrial mammals, as well as some special adaptations that enable them to survive in their watery environment.

The cetaceans, which include whales, dolphins, and porpoises, have streamlined bodies, horizontal tail flukes, and paddle-like flippers that enable them to move quickly through the water. Layers of blubber (subcutaneous fat) insulate their bodies and act as storage places for large quantities of energy. Their noses (blowholes) are located on the tops of their heads so air can be inhaled as soon as the organism surfaces above the water.

Manatees and dugongs are the only sirenians. These docile, slow-moving herbivores lack a dorsal fin or hind limbs but are equipped with front limbs that move at the elbow, as well as with a flattened tail. Their powerful tails propel them through the water, while the front limbs act as paddles for steering.

The pinnipeds—seals, sea lions, and walruses—are carnivores that have webbed feet. Although very awkward on land, the pinnipeds are agile and aggressive hunters in the water. This group of marine mammals is protected from the cold by hair and blubber. During deep-water dives, their bodies are able to restrict blood flow to vital organs and slow their heart rates to only a few beats a minute, strategies that reduce oxygen consumption. All pinnipeds come onto land or ice at breeding time.

The sea otters spend their entire lives at sea and only come ashore during storms. They are much smaller than the other marine mammals. Even though otters are very agile swimmers and divers, they are clumsy on shore. Their back feet, which are flipperlike and fully webbed, are larger than their front feet. Internally, their bodies are adapted to deal with the salt in seawater with enlarged kidneys that can eliminate the excess salt.

or have ears of reduced size, have the genitals and teats located inside the body, and have limbs that are flattened for swimming purposes.

The senses of sight and hearing are acute in pinnipeds. Their large eyes function well at low levels of light, permitting them to see in deep, murky water. They also have keen hearing, a sense that helps them locate prey. When they dive, their ears close tightly to block the entry of water. In the water their sense of smell is poor, but on land it is excellent and necessary for social interaction with other members of their species. Pinnipeds rely heavily on their sense of touch and can often be seen touching and rubbing against each other for security and reassurance.

All member of Pinnipedia spend time in water and on land. They can go for long periods of time without food and are able to stay warm in frigid conditions because their bodies store large amounts of subcutaneous fat, or blubber. Even when swimming through extremely cold ocean water, pinnipeds maintain their optimal body temperature, about 95°F (35°C), by contracting the small blood vessels to prevent large amounts of heat from escaping.

These carnivores are near the top of the ocean food chains and feed mainly on fish and squid, although they may also dine on penguins and some species of seals. Their sharp teeth are used for grasping and not chewing.

The females give birth to only one offspring each season, and they leave the water and have their babies either on land or ice. The length of mating season is determined by environmental conditions. When food is abundant and weather mild, breeding season can stretch up to eight months, but when conditions are unfavorable it may last no longer than three days.

Members of the family Otariidae, the eared seals have small external ears that are composed entirely of cartilage. Eared seals use their front flippers as support when walking, folding their hind limbs forward to improve mobility on land (see Figure 6.3). In the water, the front flippers of seals provide the main source of propulsion. Included in this group are two genera of fur seals, *Arctocephalus* and *Callorhinius*.

The eared seals live in a variety of locations including arctic, temperate, and subtropical waters. They are found along both the North and South American coasts, the coasts of central and northern Asia, and along the southwestern coast and islands of Australia. Breeding sites include sea coasts, quiet

Otariids and Phocids

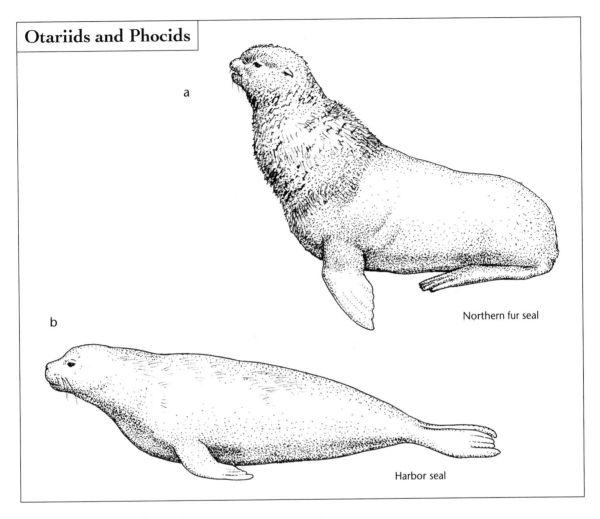

Fig. 6.3 Otariids (a), the eared seals, can bend their back limbs forward to support their bodies on land. Phocids (b), the true seals, cannot and must therefore move across the land by crawling on their ventral sides, dragging their back limbs.

bays, or rocky islands. Although their favorite foods are crustaceans, otarids will dine on other animals.

During breeding, the eared seals are extremely social animals that form large herds on land. The males, called bulls, arrive at the breeding area first to stake out their territory. Later the females show up and congregate in harems with one male to as many as 40 females. Soon after the females arrive they give birth to the pups that were conceived at last year's mating. A few days later the females are ready to mate again, beginning a pregnancy that will produce a pup next season.

The true seals, members of the Phocidae family, lack external ears although they are perfectly capable of hearing. Unlike otarids, phocids cannot fold their hind limbs forward. As a consequence, phocids move on land by hopping on their bellies while supporting themselves with their front limbs (see Figure 6.3), a technique that has earned them the nickname the "crawling seals." Although very awkward on land, phocids are more graceful in the water than most other sea mammals. In water, they are extremely agile, flipping their tails left or right to propel their bodies forward. Their skills as swimmers, coupled with their extremely streamlined shapes, allow them to perform fast, agile water maneuvers. For this reason phocids can make long trips to find food during breeding seasons, while otarids must remain close to the breeding site.

The true seals are widely distributed along coastlines above 30° north latitude and 50° south latitude. A few species are found in tropical locations, and some even reside near freshwater lakes and rivers. The young pups of true seals are often covered with a dense, soft coat, which is usually white. As the seals mature, the coat becomes short, stiff, and dark in color.

Breeding habits vary from species to species, with some true seals selecting one mate, while others opt for mating with more than one female during a breeding season. Phocids are excellent divers, and many can swim to great depths, but this ability varies from species to species. Weddell seals can reach depths of 1,968.5 feet (600 m) and stay underwater for more than an hour.

The last family of pinnipeds, Odobenidae, includes the walrus. Walrus are easily recognized because they have overgrown canines that form large tusks, as shown in the lower color insert on page C-8. These powerful teeth are used to crush mollusks and other marine invertebrates on which they feed. Walrus lack both a distinct tail as well as external ears. They lumber across the land by folding their hind limbs forward and using the front flippers for support.

Walrus live in the northern seas above the 58° latitude but are occasionally found further south. Their bodies are much larger than those of other pinnipeds, with males reaching weights of 2,645.5 pounds (1,200 kg). The thick, wrinkled skin of a walrus, which is brown with a few hairs scattered across its surface, lies atop a layer of blubber.

Congregations of walrus may include thousands of individuals. At mating time, the group subdivides, with bulls forming their own harems. In general, walrus tend to migrate as the position of the ice packs changes each season. They are excellent swimmers and divers who use their keen eyesight as well as sensory fibers around their snouts to pick up clues about the location of prey in water. Dives as deep as 328 feet (100 m) in search of crustaceans and other mollusks on the seabed are not unusual for these animals.

Conclusion

Vertebrates who live along the oceans' shores include several species of fish, birds, and mammals and one species of reptiles. The only shore reptile is the marine iguana, a lizard found only on the Galápagos Islands. Unlike terrestrial iguanas, this species is adapted to life on the edge of the sea where it feeds on intertidal algae. To dine, the lizard must dive into cold, wave-tossed ocean water and scrape the microscopic plants from rocks. Intertidal life is rough for this reptile, but its adaptations have enabled it to be successful in an environment in which no other reptile can survive.

Meanwhile, on coastlines there are innumerable birds that live in or visit the area. Shorebirds are adapted for life on land, in water, and in air with highly specialized bills, feet, and legs. One of the most common and versatile shorebirds is the seagull, a noisy black-and-white animal that can find food in almost any locale.

Shore mammals are not as numerous as birds, but their presence commands attention. Pinnipeds, a group that includes seals and walruses, are large and impressive animals. Their coastline adaptations include thick blubber for warmth, modified appendages for swimming instead of walking, and a large lung capacity that enables them to hold their breath during extended periods. Pinnipeds, like all shoreline vertebrates, endure extremely difficult day-to-day living conditions. Their superb adaptations make life in this intertidal zone not only possible but successful.

7

Change Is Constant

No part of the ocean gets more human attention than the coast. Intertidal zones, which include beaches, rocky tide pools, and mudflats, are places where people can interact with the sea and the organisms that live in it. For many, the shore is the only part of the ocean they know, but to others, it is literally just the edge of their relationship with the ocean and its bounty.

For the creatures that reside there, life on the shore is very tough. Physical and chemical factors of the shoreline are more variable than in any other part of the ocean. In addition, shorelines are unstable habitats that are constantly modified by natural forces, such as wind, waves, currents, and rain. In the space of a few hours, the temperature and salinity of shallow intertidal water can change drastically. Every organism that makes its home in the tidal zone must be superbly adapted to deal with these changes.

One of the biggest challenges to intertidal creatures is simply staying in place. The regular rush of tidal water and the endless pounding of waves mean that organisms are constantly beaten, battered, and shifted. To cope, they have developed a wide variety of adaptations. Some secure themselves to substrates with glues or fibers produced by their own bodies, while others tunnel into the sand or excavate shelters in the rocks. A few small organisms do not even attempt to stay in one place; instead, their bodies are designed to ride the ever-shifting water.

Shallow-water inhabitants are subject to a variety of predators but protect themselves with an arsenal of defense mechanisms. Sea urchins are covered in sharp spines, and anemones are armed with stinging cells. Sponges wage chemical warfare on their foes by producing toxins, and mobile animals simply run and hide, digging deep into the sand or diving under rocks for protection.

Forces That Influence the Coast

Because beaches and intertidal zones are dynamic environments, the habitats they support are constantly changing. Exposed to energy of the sea and land, they are transformed on a daily, yearly, and geologic time scale. All coasts are subject to erosion, storms, and other natural forces as well as the impact of human activities.

The continental United States has three marine shores: the Pacific coast, the Atlantic coast, and the coast of the Gulf of Mexico. Including the shorelines of the Great Lakes, U.S. coasts total more than 95,000 miles (152,855 km). All three marine coastlines are exposed to many of the same natural forces. Waves and tides continually bathe their shores while winds and precipitation rebuild their substrates. In addition, the shores in all locations are occasionally exposed to the powerful forces of storms.

Despite their similarities, the three coasts have inherent differences. The Pacific shore is the only one that is located in an area of geologic activity. There, two plates of the Earth's crust are slowly but constantly colliding, causing earthquakes and volcanic activity. The changes caused by such geologic activity are usually drastic and long lasting. Most of the sand that covers the Pacific beaches was picked up in high elevations to the east and carried to the edge of the sea by fast-moving rivers. A smaller portion of the beach sand originated from the action of the strong Pacific waves on the rocky shores.

Beaches and tidal zones along the Gulf of Mexico are protected from the full force of the ocean by the structure of the gulf. The presence of Central America blocks some of the ocean's energy, so gulf shores do not experience as much wave erosion as shores on the east and west coasts. The gulf coast is strongly influenced by the tremendous load of sediment from the Mississippi River.

For eons, the Atlantic coast has been the recipient of soil and sediment eroding from the Blue Ridge and Appalachian Mountains. As soil washed down to the ocean's border, the heavier particles dropped out first, settling along the land's margin, and the lighter ones were carried further out, to the continental shelf. Constant battering by waves has returned some of the lightweight ones to the beaches, creating wide expanses of weathered particles that are still being shifted and rearranged by the energy of the Atlantic Ocean. Storms and hurricanes often strike the Atlantic Coast and make drastic changes in its topography.

The Impact of Humans

Although the basic structures of coasts are due to the work of nature, the activities of humans change them. Coastlines make up only 11 percent of total land area of the United States but support 53 percent of its population. Where there are people, there are also cars, homes, businesses, and industries, all of which have tremendous affects on the condition of coastlines.

Most of the people at the coast do not purposefully damage the area, but the very presence of humans alters the natural environments. Some of the changes are small, and the environment can quickly recover. Vacationers exploring the sand dunes or hiking beside the marshy wetlands may tread on the plants. If foot traffic is heavy, plants that are very sensitive may die and be replaced by species that are more resistant to physical wear and tear. A change in plant life brings about a change in the structure of habitats and can eventually cause established species to disappear and new ones to show up.

Without realizing it, boaters churn waters and add petroleum compounds to them. Campers move into areas once inhabited by wild animals. Waders stir up sediment, step on underground burrows, and disturb breeding sites. Refuse from human activity means that birds get tangled in plastic lines and sea turtles ingest plastic bags, mistaking them for jellyfish.

Home construction and road building are two human endeavors that can create damaging coastal erosion. Once soil is loosened, whether by bulldozers or shovels, wind and water easily move it toward the sea's edge. There it fills in previously clear waters with sediments, reducing the availability of light in the water, covering marine plants, and filling in the homes of sediment-dwelling organisms.

Coastlines are sought-after locations for industries that ship their products along international waterways. Industry and delicate marine ecosystems rarely make good combinations because industrial pollutants can damage the coast and its native inhabitants. In some instances, poisonous waste products are dumped directly into the ocean. Oil from normal shipping practices creates hazards for the animals nearby.

People are learning to live in harmony with habitats at the sea's margin. Instead of trying to tame the coastline with seawalls and docks, many accept the natural rhythms of the Earth and water. The coasts are no longer viewed simply as places to use and exploit. Increasingly, their natural value is recognized, and they are receiving much of the respect and protection that they deserve.

Glossary

A

algal bloom The rapid growth of cyanobacteria or algae populations that results in large mats of organisms floating in the water.

amphibian A cold-blooded, soft-skinned vertebrate whose eggs hatch into larvae that metamorphose into adults.

animal An organism capable of voluntary movement that consumes food rather than manufacturing it from carbon compounds.

anterior The region of the body that is related to the front or head end of an organism.

appendage A structure that grows from the body of an organism, such as a leg or antenna.

arthropod An invertebrate animal that has a segmented body, joined appendages, and chitinous exoskeleton.

asexual reproduction A type of reproduction that employs means other than the union of an egg and sperm. Budding and binary fission are forms of asexual reproduction.

autotroph An organism that can capture energy to manufacture its own food from raw materials.

B

binary fission A type of cell division in monerans in which the parent cell separates into two identical daughter cells.

biodiversity The number and variety of life-forms that exist in a given area.

bird A warm-blooded vertebrate that is covered with feathers and reproduces by laying eggs.

bladder In macroalgae, an inflatable structure that holds gases and helps keep blades of the plant afloat.

blade The part of a nonvascular plant that is flattened and leaflike.

brood A type of behavior that enables a parent to protect eggs or offspring as they develop.

budding A type of asexual reproduction in which an offspring grows as a protrusion from the parent.

buoyancy The upward force exerted by a fluid on matter that causes the matter to tend to float.

C

carnivore An animal that feeds on the flesh of other animals.

chanocyte A flagellated cell found in the gastrovascular cavity of a sponge that moves water through the pores, into the gastrovascular cavity, and out the osculum (an exit for outflow).

chitin A tough, flexible material that forms the exoskeletons of arthropods and cell walls of fungi.

chlorophyll A green pigment, found in all photosynthetic organisms, that is able to capture the Sun's energy.

cilia A microscopic, hairlike cellular extension that can move rhythmically and may function in locomotion or in sweeping food particles toward an animal's mouth or oral opening.

cnidarian An invertebrate animal that is radially symmetrical and has a saclike internal body cavity and stinging cells.

cnidocyte A nematocyst-containing cell found in the tentacles of cnidarians that is used to immobilize prey or defend against predators.

countershading One type of protective, two-tone coloration in animals in which surfaces that are exposed to light are dark colored and those that are shaded are light colored.

cyanobacteria A moneran that contains chlorophyll as well as other accessory pigments and can carry out photosynthesis.

D

detritivore An organism that feeds on dead and decaying matter.

detritus Decaying organic matter that serves as a source of energy for detritivores.

DNA Deoxyribonucleic acid, a molecule located in the nucleus of a cell that carries the genetic information that is responsible for running that cell.

dorsal Situated on the back or upper side of an organism.

E

ecosystem A group of organisms and the environment in which they live.

endoskeleton An internal skeleton or support system such as the type found in vertebrates.

energy The ability to do work.

epidermis The outer, protective layer of cells on an organism, such as the skin.

exoskeleton In crustaceans, a hard but flexible outer covering that supports and protects the body.

F

fish A cold-blooded, aquatic vertebrate that has fins, gills, and scales and reproduces by laying eggs that are externally fertilized.

flagellum A long, whiplike cellular extension that is used for locomotion or to create currents of water within the body of an organism.

food chain The path that nutrients and energy follow as they are transferred through an ecosystem.

food web Several interrelated food chains in an ecosystem.

fungus An immobile heterotrophic organism that consumes its food by first secreting digesting enzymes on it, then absorbing the digested food molecules through the cell walls of threadlike hyphae.

G

gastrodermis The layer of cells that lines the digestive cavity of a sponge or cnidarian, and the site at which nutrient molecules are absorbed.

gastropod A class of arthropods that has either one shell or no shells, a distinct head equipped with sensory organs, and a muscular foot.

gill A structure containing thin, highly folded tissues that are rich in blood vessels and serve as the sites where gases are exchanged in aquatic organisms.

glucose A simple sugar that serves as the primary fuel in the cells of most organisms. Glucose is the product of photosynthesis.

H

herbivore An animal that feeds on plants.

hermaphrodite An animal in which both male and female sexual organs are present.

heterotroph An organism that cannot make its own food and must consume plant or animal matter to meet its body's energy needs.

holdfast The rootlike portion of a macroalga that holds the plant to the substrate.

hydrogen bond A weak bond between the positive end of one polar molecule and the negative end of another.

hyphae Filamentous strands that make up the bodies of fungi and form the threadlike extensions that produce digestive enzymes and absorb dissolved organic matter.

I

invertebrate An animal that lacks a backbone, such as a sponge, cnidarian, worm, mollusk, or arthropod.

L

lateral The region of the body that is along the side of an organism.

lateral line A line along the side of a fish that connects to pressure-sensitive nerves that enable the fish to detect vibrations in the water.

larva The newly hatched offspring of an animal that is structurally different from the adult form.

light A form of electromagnetic radiation that includes infrared, visible, ultraviolet, and X-ray that travels in waves at the speed of 186,281 miles (300,000 km) per second.

M

mammal A warm-blooded vertebrate that produces living young that are fed with milk from the mother's mammary glands.

mantle A thin tissue that lies over the organs of a gastropod and secretes the shell.

mesoglea A jellylike layer that separates the two cell layers in the bodies of sponges and cnidarians.

milt A fluid produced by male fish that contains sperm and is deposited over eggs laid by the female.

mixotroph An organism that can use the Sun's energy to make its own food or can consume food.

molt Periodic shedding of an outer layer of shell, feathers, or hair that allows new growth to occur.

moneran A simple, one-celled organism that neither contains a nucleus nor membrane-bound cell structures.

motile Capable of moving from place to place.

N

nematocyst In cnidarians, a stinging organelle that contains a long filament attached to a barbed tip that can be used in defense or to capture prey.

O

omnivore An animal that eats both plants and animals.

operculum In fish, the external covering that protects the gills. In invertebrates, a flap of tissue that can be used to close the opening in a shell, keeping the animal moist and protecting it from predators.

oviparous An animal that produces eggs that develop and hatch outside the mother's body.

ovoviviparous An animal that produces eggs that develop and hatch within the mother's body, then are extruded.

P

pectoral An anatomical feature, such as a fin, that is located on the chest.
pelvic An anatomical feature, such as a fin, that is located near the pelvis.
photosynthesis The process in which green plants use the energy of sunlight to make nutrients.
plant A nonmotile, multicellular organism that contains chlorophyll and is capable of making its own food.
polar molecule A molecule that has a negatively charged end and a positively charged end.
polychaete A member of a group of worms that has a segmented body and paired appendages.
posterior The region near the tail or hind end of an organism.
productivity The rate at which energy is used to convert carbon dioxide and other raw materials into glucose.
protist A one-celled organism that contains a nucleus and membrane-bound cell structures such as ribosomes for converting food to energy and Golgi apparati for packaging cell products.

R

radula A long muscle used for feeding that is covered with toothlike projections, found in most types of gastropods.
reptile A cold-blooded, egg-laying terrestrial vertebrate whose body is covered with scales.

S

salinity The amount of dissolved minerals in ocean water.
school A group of aquatic animals swimming together for protection or to locate food.
sessile Permanently attached to a substrate and therefore immobile.
setae Hairlike bristles that are located on the segments of polychaete worms.
sexual reproduction A type of reproduction in which egg and sperm combine to produce a zygote.
spawn The act of producing gametes, or offspring, in large numbers, often in bodies of water.
spicule In sponges, a needle-like, calcified structure located in the body wall that provides support and protection.
spiracle An opening for breathing, such as the blowhole in a whale or the opening on the head of a shark or ray.
stipe A stemlike structure in a nonvascular plant.

surface tension A measure of how easy or difficult it is for molecules of a liquid to stick together due to the attractive forces between them.
swim bladder A gas-filled organ that helps a fish control its position in the water.
symbiosis A long-term association between two different kinds of organisms that usually benefits both in some way.

T

territorial behavior The defense of a certain area or territory by an animal for the purpose of protecting food, a mate, or offspring.
thallus The body of a macroalgae, made up of the blade, stipe, and holdfast.

V

ventral Situated on the stomach or lower side of an organism.
vertebrate A member of a group of animals with backbones, including fish, amphibian, reptiles, birds, and mammals.
viviparous An animal that gives birth to living offspring.

Z

zooxanthella A one-celled organism that lives in the tissues of invertebrates such as coral, sponge, or anemone where it carries out photosynthesis.

Further Reading and Web Sites

Books

Banister, Keith, and Andrew Campbell. *The Encyclopedia of Aquatic Life.* New York: Facts On File, 1985. Well written and beautifully illustrated book on all aspects of the ocean and the organisms in it.

Coulombe, Deborah A. *The Seaside Naturalist.* New York: Fireside, 1990. A delightful book for young students who are beginning their study of ocean life.

Davis, Richard A. *Oceanography: An Introduction to the Marine Environment.* Dubuque, Iowa: Wm. C. Brown Publishers, 1991. A text that helps students become familiar with and appreciate the world's oceans.

Dean, Cornelia. *Against the Tide.* New York: Columbia University Press, 1999. An analysis of the impact of humans and nature on the ever-changing beaches.

Ellis, Richard. *Encyclopedia of the Sea.* New York: Alfred A. Knopf, 2000. A factual, yet entertaining, compendium of sea life and lore.

Garrison, Tom. *Oceanography.* New York: Wadsworth Publishing, 1996. An interdisciplinary examination of the ocean for beginning marine science students.

Karleskint, George, Jr. *Introduction to Marine Biology.* Belmont, Calif.: Brooks/Cole-Thompson Learning, 1998. An enjoyable text on marine organisms and their relationships with one another and with their physical environments.

McCutcheon, Scott, and Bobbi McCutcheon. *The Facts On File Marine Science Handbook.* New York: Facts On File, 2003. An excellent resource that includes information on marine physical factors and living things as well as the people who have been important in ocean studies.

Nowak, Ronald M., et al. *Walker's Marine Mammals of the World.* Baltimore, Md.: Johns Hopkins University Press, 2003. An overview on the anatomy, taxonomy, and natural history of the marine mammals.

Pinet, Paul R. *Invitation to Oceanography.* Sudbury, Mass.: Jones and Bartlett Publishers, 2000. Includes explanations of the causes and effects of tides and currents, as well as the origins of ocean habitats.

Prager, Ellen J. *The Sea.* New York: McGraw-Hill, 2000. An evolutionary view of life in the Earth's oceans.

Reeves, Randall R., et al. *Guide to Marine Mammals of the World.* New York: Alfred A. Knopf, 2002. An encyclopedic work on sea mammals accompanied with gorgeous color plates.

Rice, Tony. *Deep Oceans*. Washington, D.C.: Smithsonian Museum Press, 2000. A visually stunning look at life in the deep ocean.

Sverdrup, Keith A., Alyn C. Duxbury, and Alison B. Duxbury. *An Introduction to the World's Oceans*. New York: McGraw Hill, 2003. A comprehensive text on all aspects of the physical ocean, including the seafloor and the ocean's physical properties.

Thomas, David. *Seaweeds*. Washington, D.C.: Smithsonian Museum Press, 2002. Illustrates and describes seaweeds from microscopic forms to giant kelps, explaining how they live, what they look like, and why humans value them.

Thorne-Miller, Boyce, and John G. Catena. *The Living Ocean*. Washington, D.C.: Friends of the Earth, 1991. A study of the loss of diversity in ocean habitats.

Waller, Geoffrey. *SeaLife: A Complete Guide to the Marine Environment*. Washington, D.C.: Smithsonian Institution Press, 1996. A text that describes the astonishing diversity of organisms in the sea.

Web Sites

Bird, Jonathon. *Adaptations for Survival in the Sea*, Oceanic Research Group, 1996. Available online. URL: http://www.oceanicresearch.org/adapspt.html. Accessed March 19, 2004. A summary and review of the educational film of the same name, which describes and illustrates some of the adaptations that animals have for life in salt water.

Buchheim, Jason. "A Quick Course in Ichthyology." Odyssey Expeditions. Available online. URL: http://www.marinebiology.org/fish.htm. Accessed January 4, 2004. A detailed explanation of fish physiology.

"Conservation: Why Care About Reefs?" REN Reef Education Network, Environment Australia. Available online. URL: http://www.reef.edu.au/asp_pages/search.asp. Accessed November 18, 2004. A superb Web site dedicated to the organisms living in and the health of the coral reefs.

Duffy, J. Emmett. "Underwater urbanites: Sponge-dwelling napping shrimps are the only known marine animals to live in colonies that resemble the societies of bees and wasps." *Natural History*. December 2003. Available online. URL: http://www.findarticles.com/cf_dls/m1134/10_111736243/print.jhtml. Accessed January 2, 2004. A readable and fascinating explanation of eusocial behavior in shrimp and other animals.

"Fungus Farming in a Snail." *Proceedings of the National Academy of Science,* 100, no. 26 (December 4, 2003). Available online. URL: http://www.pnas.org/cgi/content/abstract/100/26/15643. A well-written, in-depth analysis of the ways that snails encourage the growth of fungi for their own food.

Gulf of Maine Research Institute Web site. Available online. URL: http://www.gma.org/about_GMA/default.asp. Accessed January 2, 2004. A comprehensive and up-to-date research site on all forms of marine life.

"Habitat Guides: Beaches and Shorelines." eNature. Available online. URL: http://www.enature.com/habitats/show_sublifezone.asp?sublifezoneID=60#Anchor-habitat-49575. Accessed November 21, 2003. A Web site with young people in mind that provides comprehensive information on habitats, organisms, and physical ocean factors.

Huber, Brian T. "Climate Change Records from the Oceans: Fossil Foraminifera." Smithsonian National Museum of Natural History. June 1993. Available online. URL: http://www.nmnh.si.edu/paleo/marine/foraminifera.htm. Accessed December 30, 2003. A concise look at the natural history of foraminifera.

"Index of Factsheets." Defenders of Wildlife. Available online. URL: http://www.kidsplanet.org/factsheets. Accessed November 18, 2004. Various species of marine animals are described on this excellent Web site suitable for both children and young adults.

King County's Marine Waters Web site. Available online. URL: http://splash.metrokc.gov/wlr/waterres/marine/index.htm. Accessed December 2, 2003. A terrific Web site on all aspects of the ocean, emphasizing the organisms that live there.

Mapes, Jennifer. "U.N. Scientists Warn of Catastrophic Climate Changes." National Geographic News. February 6, 2001. Available online. URL: http://news.nationalgeographic.com/news/2001/02/0206_climate1.html. A first-rate overview of the current data and consequences of global warming.

National Oceanic and Atmospheric Administration Web site. Available online. URL: http://www.noaa.gov/. A top-notch resource for news, research, diagrams, and photographs relating to the oceans, coasts, weather, climate, and research.

"Resource Guide, Elementary and Middle School Resources: Physical Parameters." Consortium for Oceanographic Activities for Students and Teachers. Available online. URL: http://www.coast-nopp.org/toc.html. Accessed December 10, 2003. A Web site for students and teachers that includes information and activities.

"Sea Snakes in Australian Waters." CRC Reef Research Centre. Available online. URL: http://www.reef.crc.org.au/discover/plantsanimals/seasnakes. Accessed November 18, 2004. An overview of sea snake classification, breeding, and venom.

U.S. Fish and Wildlife Service Web site. Available online. URL: http://www.fws.gov/. A federal conservation organization that covers a wide range of topics, including fisheries, endangered animals, the condition of the oceans, and conservation news.

Index

Note: *Italic* page numbers indicate illustrations.
C indicates color insert pages.

A

acorn barnacles 72, 73
aggregate green anemone (*Anthopleura elegantissima*) 50
algae *See* macroalgae
alginate 37
alternation of generations 35
amoebocytes 43, 43–44
amphipods 72
anemones 48–51, *51*, C-1
Animalia kingdom 27
arthropods 69–77
    crustaceans 71–76
    horseshoe crabs 77, C-5
    sea spiders 76–77
asexual reproduction 29–30
Atlantic coast 116
autotrophs (producers) 23

B

backwash marks 9
bacteria 24–28
bamboo worm (*Clymenella*) 59
barnacles 71–72
barrier islands 9
beaches 8–9, 14–15
berms 8–9
bilateral symmetry 46, *47*
bills 104, 105, *105*
binary fission 29–30
birds 103–108
    anatomy 104, *105*
    gulls (Laridae) 107–108, *108*
    plovers (Charadriidae) 106–107
    sandpipers (Scolopacidae) 106, C-8
bivalves 63, 63–64, 67–69
black zone 25
bladders 34, 35
blades *34*, 35
blennies (rock skippers, Blenniidae) 90–91
bloodworms (*Glycera*) 56
blue crab (*Callinectes sapidus*) 74
body symmetry 46, *47*
body temperature 98–99
boring sponges (*Cliona*) 45
breadcrumb sponge (*Halichondria panicea*) 45
brittle stars 79
brooding 52
brown algae 36–38
butterfish (*Pholis gunnellus*) 95

C

Calothrix 25
carbon dioxide 2, 3, 11 12, 13
cetaceans 109
chitons 64, C-3
choanocytes 43, 44
ciliates 31–32
clams 63, 68
clam worms (*Nereis*) 58
clingfish (Gobiesocidae) 95–96
cnidarians 48–53
    anemones 48–51, *51*, C-1
    hydroids 51–53
    jellyfish 53
    octocorals 53
coasts
    Atlantic coast 116
    beaches and sand 8–9, 14–15, *C-1*
    characteristics of 4–5
    formation of 5–8
    Gulf of Mexico 115
    habitats 19–20
    high energy *vs.* low energy 8
    human impact on 116–117
    Pacific coast 115
    primary coasts 5–7
    secondary coasts 7–8
    substrates 14–15, 22
    water characteristics 10–14
    water processes 7–8, 15–19
cockles 69
Codium fragile 36
cold-blooded animals (ectotherms) 99
coloration 33, 88–89
colored sand (Farbstreifensandwatt) 28
common blenny (*Lipophrys pholis*) 90–91

conchs 66–67
consumers (heterotrophs) 23
coralline algae 39
corals 9
crabs 74–76, 75, 76
crocodiles 101
crustaceans 71–76
currents 18–19
cusps 9
cyanobacteria 24–26

D

damselfish 96
dead man's fingers (*Alcyonium digitatum*) 53
dead man's fingers (*Codium fragile*) 36
decomposers (detritivores) 23
defensive mechanisms 38, 88–89
deltas 6
density 13–14
detritivores (decomposers) 23
diatoms 28–31, *31*
dinoflagellates 49–50
dolphins 109
drowned rivers 6
dugongs 109

E

eared seals (Otariidae) 110–111, *111*
echinoderms 77–81
ectotherms (cold-blooded animals) 99
endotherms (warm-blooded animals) 98–99
Enteromorpha intestinalis 36
erosion 7–9
errant species 55
exoskeletons 70

F

fault bays 7
feather hydroids 52–53
finger sponge (*Haliclona oculata*) 45
fins 86, 87
fish 83–97
  anatomy 86–87
  blennies (rock skippers, Blennidae) 90–91
  clingfish (Gobiesocidae) 95–96
  color 88–89
  ectotherms (cold-blooded animals) 99
  gobies (goby, Gobiidae) 88–89
  gunnels (Pholidae) 92–95
  sculpins (Cottidae) 83–85, *84*, 88
  senses 92–93
  skin 92
  territorial behavior 96
  water balance 94

flagella *43*, 44
flatworms 53–55, 56–57
food chains 23
frustules 28–29, *29*
Fucus (rockweed) 36–37
fungi 27, 32
fur seals 110–111, *111*

G

gases, dissolved in seawater 11–13
gastropods 64–67
ghost crabs (*Ocypode quadrata*) 75, 76
giant green anemone 50–51
gills 62
glaciers 6–7
global warming 2
glucose 23
gobies (goby, Gobiidae) 88–89
goose barnacles 72
grasses 39–40
green algae 33, 35–36, 50
green crab (*Carcinus maenas*) 74–75
greenhouse gases 2, 3
Gulf of Mexico 115
gulls (Laridae) 107–108, *108*
gunnels (Pholidae) 92–95

H

habitats 19–20
hermit crabs 75–76
herring gull (*Larus argentatus*) 108, *108*
heterotrophs (consumers) 23
holdfasts *34*, 35, 36
horseshoe crabs 77, C-5
hydroids 51–53

I

ice ages 4–5, 6
iguana (*Amblyrhynchus cristatus*) 99–103, C-7
intertidal (littoral) zone 19–20, 22, 24
invertebrates
  arthropods 69–77
  cnidarians 48–53
  complex, characteristics of 62–63
  echinoderms 77–81
  exoskeletons 70
  mollusks 63–69
  simple, characteristics of 42
  sponges 42–45, *43*
  worms 53–59
Irish moss (*Chondrus crispus*) 39
isopods 72

Index 131

J
jellyfish 53

K
kelp *40*
keyhole limpets 65
kingdom classification system 26–27

L
limpets 64–65
Linnaeus, Carolus 26
littoral (intertidal) zone 19–20, 22, 24
lobsters 74, C-5
long-eyed shrimp (*Ogyrides alphaerostris*) 73
lugworm (*Arenicola*) 59
Lyngbya aestuarii 25–26

M
macroalgae (seaweeds) 32–39
    brown algae 36–38
    coloration 33
    green algae 33, 35–36, 50
    parts of *34*, 35
    red algae 38–39
mammals 108–113
    eared seals (Otariidae) 110–111, *111*
    endotherms 98–99
    true seals (Phocidae) *111*, 111–112
    walrus (Odobenidae) 112–113, C-8
manatees 109
mangroves 10
marine iguana (*Amblyrhynchus cristatus*) 99–103, C-7
mesoglea 43, *43*
milky ribbon worm (*Cerebratulus lacteus*) 54–55
mollusks 63–69
    bivalves *63*, 63–64, 67–69
    chitons 64, C-3
    gastropods 64–67
monerans 24–28
moon snails 65
moraines 6–7
mud flats 15
Murex 65
mussels 68

N
neap tides 16, *17*
nudibranchs (sea slugs) 64, 67, C-3, C-4

O
ocean currents 18–19
ocean levels 4–5
octocorals 53
orbits *15*, 17
Oregon coast *1*, 4
organpipe sponges (*Leucosolenia*) 45
oscula (osculum) *43*, 44
oxygen 11–12, 13
oystercatchers (Haematopodidae) 105–106
oyster flatworm (*Stylochus ellipticus*) 54
oysters 69, C-4

P
Pacific coast 115
parapodia (parapodium) 55
parchment worm (*Chaetopterus variopedatus*) 59
pennarians (*Pennaria*) 52–53
penpoint gunnel (*Apodichthys flavidus*) 94–94
periwinkles 66
Phocidae (true seals) *111*, 111–112
photosynthesis 23, 33
pinnipeds (Pinnipedia) 108–113
plants (Plantae) 27, 32–40
    coloration 33
    defensive mechanisms 38
    differences in terrestrial and aquatic 34–45
    macroalgae (seaweeds) 32–39, *34*
    sea grasses 39–40, *40*
    types of 32, 34–35
plovers (Charadriidae) 106–107
polychaetes (segmented worms) 55–59
porpoises 109
primary coasts 5–7
producers (autotrophs) 23
protists (Protista) 27, 28–32
    ciliates 31–32
    diatoms 28–31, *31*
purple laver (*Porphyra capensis*) 39
purple sponge (*Haliclona permollis*) 45

R
red algae 38–39
reptiles 99–103
    anatomy 100
    ectotherms 99
    marine iguana (*Amblyrhynchus cristatus*) 99–103, C-7
rivers 6
rock skippers (Blennidae) 90–91

S
sabellid worm 58, C-2
salinity (salts) 10–11, 13–14